Eco-Friendly Textile Dyeing and Finishing

Eco-Friendly Textile Dyeing and Finishing

Editor

Jamshed A Khan

scitus
academics

Eco-Friendly Textile Dyeing and Finishing
Edited by **Jamshed A Khan**

ISBN: 978-1-68117-323-8
Library of Congress Control Number: 2016934778

SCITUS Academics LLC,
www.scitusacademics.com
Box No. 4766, 616 Corporate Way,
Suite 2, Valley Cottage,
NY 10989

Printed in United States of America on Acid Free Paper

Preface

Years of human ignorance has diminished our natural resources and aged our planet. Now, people are making an effort to change the way they are treating the planet. Being more environmentally conscious about the impact materials used for fashion have on our planet is one-way designers can reduce waste and help enable a better world. By going eco-friendly can be less harmful to our natural resources. Not all fashion is following this eco-friendly trend, but more designers are embracing the trend toward eco-fashion than ever before. If the entire fashion industry became eco-friendly, it would make a huge difference for future generations because the fashion industry employs over a billion people globally. There is need for eco-friendly wet processing that is sustainable and beneficial methods. Number of sustainable practices has been implemented by various textile processing industries such as Eco- friendly bleaching; Peroxide bleaching; Eco-friendly dyeing and Printing; Low impact dyes; Natural dyes; Azo Free dyes; Phthalates Free Printing. There are a variety of materials considered "environmentally-friendly" for a variety of reasons. The industry is desperately in the need of newer and very efficient dyeing/finishing and functional treatments of textiles. There is growing awareness and readiness to adapt new perspective on industrial upgradation of Cleaner Production Programme, such new technologies help enterprises achieve green production and cost reduction at the same time. Green Production has become necessary for enterprises under the upgrade and transformation policy. The book Eco-Friendly Textile Dyeing and Finishing covers topics in the area of sustainable practices in textile dyeing and finishing.

Table of Contents

CHAPTER 1

Textile Dyes: Dyeing Process and Environmental Impact

Farah Maria Drumond Chequer[1], Gisele Augusto Rodrigues de Oliveira[1], Elisa Raquel Anastácio Ferraz[1], Juliano Carvalho Cardoso[2], Maria Valnice Boldrin Zanoni[2] and Danielle Palma de Oliveira[1]

[1] USP, Department of Clinical, Toxicological and Bromatological Analyses, Faculty of Pharmaceutical Sciences at Ribeirão Preto, University of São Paulo, Ribeirão Preto, SP, Brazil

[2] UNESP, Department of Analytical Chemistry, Institute of Chemistry at Araraquara, State of São Paulo University Júlio de Mesquita Filho, Araraquara, SP, Brazil

1. INTRODUCTION

Dyes may be defined as substances that, when applied to a substrate provide color by a process that alters, at least temporarily, any crystal structure of the colored substances [1,2]. Such substances with considerable coloring capacity are widely employed in the textile, pharmaceutical, food, cosmetics, plastics, photographic and paper industries [3,4]. The dyes can adhere to compatible surfaces by solution, by forming covalent bond or complexes with salts or metals, by physical adsorption or by mechanical retention [1,2]. Dyes are classified according to their application and chemical structure, and are composed of a group of atoms known as chromophores, responsible for the dye color. These chromophore-containing centers are based on diverse functional groups, such as azo, anthraquinone, methine, nitro, arilmethane, carbonyl and others. In addition, electrons withdrawing or donating substituents so as to generate or intensify the color of the chromophores are denominated as auxochromes. The most common auxochromes are amine, carboxyl, sulfonate and hydroxyl [5-7].

It is estimated that over 10,000 different dyes and pigments are used industrially and over 7 x 10^5 tons of synthetic dyes are annually produced worldwide [3,8,9]. Textile materials can be dyed using batch, continuous or semi-continuous processes. The kind of process used depends on many characteristics including type of material as such fiber, yarn, fabric, fabric construction and garment, as also the generic type of fiber, size of dye lots and quality requirements in the dyed fabric. Among these processes, the batch process is the most common method used to dye textile materials [10].

In the textile industry, up to 200,000 tons of these dyes are lost to effluents every year during the dyeing and finishing operations, due to the inefficiency of the dyeing process [9]. Unfortunately, most of these dyes escape conventional wastewater treatment processes and persist in the environment as a result of their high stability to light, temperature, water, detergents, chemicals, soap and other parameters such as bleach and perspiration [11]. In addition, anti-microbial agents resistant to biological degradation are frequently used in the manufacture of textiles, particularly for natural fibers such as cotton [11,12]. The synthetic origin and complex aromatic structure of these agents make them more recalcitrant to biodegradation [13,14]. However, environmental legislation obliges industries to eliminate color from their dye-containing effluents, before disposal into water bodies [9, 12].

The textile industry consumes a substantial amount of water in its manufacturing processes used mainly in the dyeing and finishing operations of the plants. The wastewater from textile plants is classified as the most polluting of all the industrial sectors, considering the volume generated as well as the effluent composition [15-17]. In addition, the increased demand for textile products and the proportional increase in their production, and the use of synthetic dyes have together contributed to dye wastewater becoming one of the substantial sources of severe pollution problems in current times [6,9].

Textile wastewaters are characterized by extreme fluctuations in many parameters such as chemical oxygen demand (COD), biochemical oxygen demand (BOD), pH, color and salinity. The composition of the wastewater will depend on the different organic-based compounds, chemicals and dyes used in the dry and wet-processing steps [6,18]. Recalcitrant organic, colored, toxicant, surfactant and chlorinated compounds and salts are the main pollutants in textile effluents [17].

In addition, the effects caused by other pollutants in textile wastewater, and the presence of very small amounts of dyes (<1 mg/L for some dyes) in the water, which are nevertheless highly visible, seriously affects the aesthetic quality and transparency of water bodies such as lakes, rivers and others, leading to damage to the aquatic environment [19,20].

During the dyeing process it has been estimated that the losses of colorants to the environment can reach 10–50% [13,14,17,21,22]. It is noteworthy that some dyes are highly toxic and mutagenic, and also decrease light penetration and photosynthetic activity, causing oxygen deficiency and limiting downstream beneficial uses such as recreation, drinking water and irrigation [13,14,23]

With respect to the number and production volumes, azo dyes are the largest group of colorants, constituting 60-70% of all organic dyes produced in the world [2,24]. The success of azo dyes is due to the their ease and cost effectiveness for synthesis as compared to natural dyes, and also their great structural diversity, high molar extinction coefficient, and medium-to-high fastness properties in relation to light as well as to wetness [2,25]. They have a wide range of applications in the textile, pharmaceutical and cosmetic industries, and are also used in food, paper, leather and paints [26,27]. However, some azo dyes can show toxic effects, especially carcinogenic and mutagenic events [27,28].

The toxic effects of the azo dyes may result from the direct action of the agent itself or of the aryl amine derivatives generated during reductive biotransformation of the azo bond [22]. The azo dyes entering the body by ingestion can be metabolized to aromatic amines by the azoreductases of intestinal microorganisms. If the dyes are nitro, they can be metabolized by the nitroredutases produced by the same microorganisms [29]. Mammalian liver enzymes and other organizations may also catalyze the reductive cleavage of the azo bond and the nitroreduction of the nitro group. In both cases, if N-hydroxylamines are formed, these compounds are capable of causing DNA damage [29, 30].

One of the most difficult tasks confronted by the wastewater treatment plants of textile industries is the removal of the color of these compounds, mainly because dyes and pigments are designed to resist biodegradation, such that they remain in the environment for a long period of time. For example, the half-life of the hydrolyzed dye Reactive Blue 19 is about 46 years at pH 7 and 25°C [31,32].

Carneiro et al. (2010) designed and optimized an accurate and sensitive analytical method for monitoring the dyes C.I. Disperse Blue 373 (DB373), C.I. Disperse Orange 37 (DO37) and C.I. Disperse Violet 93 (DV93) in environmental samples. This investigation showed that DB373, DO37 and DV93 were present in both untreated river water and drinking water, indicating that the effluent treatment (pre-chlorination, flocculation, coagulation and flotation) generally used by drinking water treatment plants, was not entirely effective in removing these dyes. This study was confirmed by the mutagenic activity detected in these wastewaters [33].

In this context, and considering the importance of colored products in present day societies, it is of relevance to optimize the coloring process with the objective of reducing the environmental impact of the textile industry. For this purpose, liposomes could be used to carry several encapsulated dyes, and hence improve the mechanical properties of textile products, resulting in better wash fastness properties and reducing the process temperature, thus economizing energy [34]. Another way is to use ultrasonic energy, studied with the objectives of improving dye productivity and washing fastness, and reducing both energy costs and water consumption [35].

Considering the fact that the textile dyeing process is recognized as one of the most environmentally unfriendly industrial processes, it is of extreme importance to understand the critical points of the dyeing process so as to find alternative, eco-friendly methods.

2. DYEING PROCESS

The dyeing process is one of the key factors in the successful trading of textile products. In addition to the design and beautiful color, the consumer usually looks for some basic product characteristics, such as good fixation with respect to light, perspiration and washing, both initially and after prolonged use. To ensure these properties, the substances that give color to the fiber must show high affinity, uniform color, resistance to fading, and be economically feasible [36].

Modern dyeing technology consists of several steps selected according to the nature of the fiber and properties of the dyes and pigments for use in fabrics, such as chemical structure, classification, commercial availability, fixing properties compatible with the target material to be dyed, economic considerations and many others [36].

Dyeing methods have not changed much with time. Basically water is used to clean, dye and apply auxiliary chemicals to the fabrics, and also to rinse the treated fibers or fabrics [37]. The dyeing process involves three steps: preparation, dyeing and finishing, as follows:

Preparation is the step in which unwanted impurities are removed from the fabrics before dyeing. This can be carried out by cleaning with aqueous alkaline substances and detergents or by applying enzymes. Many fabrics are bleached with hydrogen peroxide or chlorine-containing compounds in order to remove their natural color, and if the fabric is to be sold white and not dyed, optical brightening agents are added [37].

Dyeing is the aqueous application of color to the textile substrates, mainly using synthetic organic dyes and frequently at elevated temperatures and pressures in some of the steps [37,38]. It is important to point out that there is no dye which dyes all existing fibers and no fiber which can be dyed by all known dyes [39]. During this step, the dyes and chemical aids such as surfactants, acids, alkali/bases, electrolytes, carriers, leveling agents, promoting agents, chelating agents, emulsifying oils, softening agents etc [23,38] are applied to the textile to get a uniform depth of color with the color fastness properties suitable for the end use of the fabric [37].This process includes diffusion of the dye into the liquid phase followed by adsorption onto the outer surface of the fibers, and finally diffusion and adsorption on the inner surface of the fibers [40]. Depending on the expected end use of the fabrics, different fastness properties may be required. For instance, swimsuits must not bleed in water and automotive fabrics should not fade after prolonged exposure to sunlight [37]. Different types of dye and chemical additives are used to obtain these properties, which is carried out during the finishing step. Dyeing can also be accomplished by applying pigments (pigments differ from dyes by not showing chemical or physical affinity for the fibers) together with binders (polymers which fix the pigment to the fibers) [39,41].

Finishing involves treatments with chemical compounds aimed at improving the quality of the fabric. Permanent press treatments, water proofing, softening, antistatic protection, soil resistance, stain release and microbial/fungal protection are all examples of fabric treatments applied in the finishing process [37].

Dyeing can be carried out as a continuous or batch process [37]. The most appropriate process to use depends on several factors, such as type of material (fiber, yarn, fabric, fabric construction, garment), generic type of fiber, size of dye batch and quality requirements for the dyed fabric, but batch processes are more commonly used to dye textile materials [10].

In continuous processing, heat and steam are applied to long rolls of fabric as they pass through a series of concentrated chemical solutions. The fabric retains the greater part of the chemicals while rinsing removes most of the preparation chemicals. Each time a fabric is passed through a solution, an amount of water equivalent to the weight of the fabric must be used [37].

In batch processing, sometimes called exhaust dyeing, since the dye is gradually transferred from the dye bath to the material being dyed over a relatively long period of time [10], the dyeing occurs in the presence of dilute chemicals in a closed equipment such as a kier, kettle, beam, jet or beck [37]. Unlike the continuous process, instead of being passed through various baths in a long series of equipment sections, in the batch process the fabric remains in a single piece of equipment, which is alternately filled with water and then drained, at each step of the process. Each time the fabric is exposed to a separate bath, it uses five to ten times its own weight in water [37].

Some batch dyeing machines only operate at temperatures up to 100ºC. However, the system can be pressurized, allowing for the use of temperatures above 100ºC. Cotton, rayon, nylon, wool and some other fibers dye well at temperatures of 100ºC or below. Polyester and some other synthetic fibers dye more easily at temperatures above 100ºC [10].

Since the degree of dye fixation depends on the nature of the fiber, it is important to consider this topic. The fibers used in the textile industry can be divided into two main groups denominated natural fibers and synthetic fibers [36,37]. Natural fibers are derived from the environment (plants or animals), such as wool, cotton, flax, silk, jute, hemp and sisal, most of which are based on cellulose and proteins. On the other hand, synthetic fibers are organic polymers, mostly derived from petroleum sources, for example, polyester, polyamide, rayon, acetate and acrylic [37,39,42]. The two most important textile fibers are cotton, the largest, and polyester [43,44].

Cotton has been used for over 7000 years, and consists of mainly cellulose, natural waxes and proteins. The large number of hydroxyl groups on the cellulose provides a great water absorption capacity [39].

Several aromatic polyesters have been synthesized and studied. Of these, polyethylene terephthalate (PET) and polybutylene terephthalate (PBT) have been produced commercially for more than 50 years. Amongst other uses, PET has been used worldwide for the production of synthetic fibers due to its good physical properties. PET is manufactured from ethylene glycol (EG) and terephthalic acid (TPA) or dimethyl terephthalate (DMT). The polymerization proceeds in two steps: esterification and condensation reactions [45]. The polymer produced after condensation is solidified by jets of cold water and cut into regular granules which frequently have a cubic form. Then, the polymer melt is spun and the yarns are solidified by a stream of cold air [39].

The dye can be fixed to the fiber by several mechanisms, generally in aqueous solution, and may involve primarily four types of interaction: ionic, Van der Waals and hydrogen interactions, and covalent bonds [5].

Ionic interactions result from interactions between oppositely charged ions present in the dyes and fibers, such as those between the positive center of the amino groups and carboxyl groups in the fiber and ionic charges on the dye molecule, and the ionic attraction between dye cations and anionic groups ($-SO_3^-$ and $-CO_2^-$) present in the acrylic fiber polymer molecules. Typical examples of this type of interaction can be found in the dyeing of wool, silk and polyamide [5,36,41].

Van der Waals interactions come from a close approach between the π orbitals of the dye molecule and the fiber, so that the dye molecules are firmly "anchored" to the fiber by an affinity process without forming an actual bond. Typical examples of this type of interaction are found in the dyeing of wool and polyester with dyes with a high affinity for cellulose [36].

Hydrogen interactions are formed between hydrogen atoms covalently bonded in the dye and free electron pairs of donor atoms in the center of the fiber. This interaction can be found in the dyeing of wool, silk and synthetic fibers such as ethyl cellulose [5,36].

Covalent bonds are formed between dye molecules containing reactive groups (electrophilic groups) and nucleophilic groups on the fiber, for example, the bond between a carbon atom of the reactive dye molecule and an oxygen, nitrogen or sulfur atom of a hydroxy, amino or thiol group present in the textile fiber. This type of bond can be found in the dyeing of cotton fiber [5,36,46].

2.1. Application of Liposome-Based Technology in Textile Dyeing Process

There is increasing interest in the textile industry in the development of eco-friendly textile processing, in which the use of naturally occurring materials such as phospholipids, would become important [47]. Phospholipids are natural surfactants and in the presence of water, they organize themselves so as to reduce unfavorable interactions between their hydrophobic tails and the aqueous solution; their hydrophilic head groups exposed to the aqueous phase forming vesicles. Liposomes or phospholipid vesicles are featured by clearly separate hydrophilic and hydrophobic regions [34,48].

Liposomes were first produced in England in 1961 by Alec D. Bangham, who was studying phospholipids and blood clotting. He found that when phospholipids were added to water, they immediately formed a sphere, because one end of each molecule was water soluble, while the opposite end was water insoluble [49]. From a chemical point of view, the liposome is an amphoteric compound containing both positive and negative charges [34,50].

Liposomes are defined as a structure composed of lipid vesicle bilayers which can encapsulate hydrophobic or hydrophilic compounds in the lipid bilayer or in the aqueous volume, respectively [51]. These structures are usually made up of phosphatidylcholine (PC), which has a hydrophilic part consisting of phosphate and choline groups and a hydrophobic part composed of two hydrocarbon chains of variable length [49, 52,53].

Liposomes are often distinguished according to their number of lamellae and size. Small unilamellar vesicles (SUV), large unilamellar vesicles (LUV) and large multilamellar vesicles (MLV) or multivesicular vesicles (MVV) can be differentiated [49]. The diameters of liposomes vary from a nanometer to a micrometer [34]. Multilamellar liposomes (MLV) usually range from 500 to 10,000 nm. Unilamellar liposomes can be small (SUV) or large (LUV); SUV are usually smaller than 50 nm and LUV are usually larger than 50 nm. Very large liposomes are called giant liposomes (10,000 - 10,00,000 nm). They can be either unilamellar or multilamellar. The liposomes containing encapsulated vesicles are called multi-vesicular and they range from 2,000-40,000 nm. LUVs with an asymmetric distribution of phospholipids in the bilayers are called asymmetric liposomes [54]. The thickness of the membrane (phospholipid bilayer) measures approximately 5 to 6 nm [49].

According to Sivasankar, Katyayani (2011), the preparation of liposomes is based on lipids, and those normally used are:

Natural phospholipids:

- Phosphatidyl choline (PC) - Lecithin
- Phosphatidyl ethanolamine (PE) - Cephalin
- Phosphatidyl serine (PS)
- Phosphatidyl inositol (PI)
- Phosphatidyl glycerol (PG)

Synthetic phospholipids:

For saturated phospholipids:

- Dipalmitoyl phosphatidyl choline (DPPC)
- Distearoyl phosphatidyl choline (DSPC)
- Dipalmitoyl phosphatidyl ethanolamine (DPPE)
- Dipalmitoyl phosphatidyl serine (DPPS)
- Dipalmitoyl phosphatidic acid (DPPA)
- Dipalmitoyl phosphatidyl glycerol (DPPG)

For unsaturated phospholipids:

- Dioleoyl phosphatidyl choline (DOPC)
- Dioleoyl phosphatidyl glycerol (DOPG) [52].

Natural acidic lipids, such as PS, PG, PI, PA (phosphatidic acid) and cardiolipin (CL), are added when anionic liposomes are desired, and cholesterol is often included to stabilize the bilayer. These molecules are derivatives of glycerol with two alkyl groups and one amphoteric group [34].

Phosphatidylcholine is the biological lipid most widely used for producing liposomes. Liposomes based on phosphatidylcholine consist of phosphatidic acid and glycerin, with two alcohol groups esterified by fatty acids and a third group esterified by phosphoric acid, to which the amino alcohol choline is added as a polar group [34,49].

According to Barani, Montazer (2008), normally four different methods can be used for the preparation of liposomes:

1. Dry lipid film;
2. Emulsions;
3. Micelle-forming detergents;
4. Alcohol injection technology [34].

Liposomes have two distinct roles: they can provide an excellent model for biological membranes, and they are being developed as controlled delivery systems for hydrophilic and lipophilic agents [34,53,55]. They are promising candidates for adjuvant and carrier systems for drug delivery, are well-documented, and can be used for the same purpose in textile materials [33].

Encapsulation or liposome technology is applied in numerous fields, such as in pharmaceuticals, cosmetics, foods, detergents, textiles and other applications where it is important to liberate the encapsulated material slowly [34,50]. This new clean technology has already been adopted by some textile industries [52]. In recent years, liposomes have been examined as a way of delivering dyes to textiles in a cost-effective and environmentally sensitive way [56].

Conventional dyeing processes consume a great deal of energy, a significant amount of which is wasted in controlling the process parameters in order to achieve uniform results. With respect to the carrier role of liposomes, they can be used in several textile processes such as textile finishing and dyeing, with several types of dyes and fibers. They are nontoxic, biodegradable, and can encapsulate a wide range of solutes [34]. In addition, the main advantages of liposomes are a clear reduction in dyeing temperature (about 10°C as compared to conventional dyeing), improved quality of the textiles produced, with additional benefits with respect to material weight yield during subsequent spinning, improved smoothness and mechanical properties of the dyed textiles, and a clear reduction in the contamination load of the dye baths [52,57]. Low temperature gives a more natural feel and improved quality, with lower environmental impact [34].

In recent years, liposomes have been used in the textile industry as a carrier for auxiliary materials (leveling, retarding and wetting agents) in dyeing, mainly for wool dyeing, and for finishing processes [34,55]. One of the most common problems with textile auxiliaries is that they fail to form a complex in the solution bath. This problem can be solved by using liposomes with selected positive or negative charges. Liposomes can be prepared according to the type of process, solute material and fiber structure [34]. Liposomes from phospholipids have been widely used as a dye carrier in the dyeing process, and create eco-friendly textile processes. Due to their structural properties, liposomes can encapsulate hydrophilic dyes (reactive, acid and basic dyes) in the aqueous phase, and hydrophobic dyes (disperse dyes) in the phospholipid bilayers [58]. Liposomes containing a dye are generally large, irregular and unilamellar [50].

According to Barani & Montazer (2008), the application of liposomes in textile processing can be useful when the release of the solute material is important, and improves the final properties of the products. A wetting agent is required in the conventional bleaching bath of cotton fabrics, but this step can be eliminated by using liposomes. The presence of liposomes in the peroxide bleaching bath can improve the mechanical properties of fabrics and their brightness. Liposomes contain particles of oxidant present in the bleaching solution that represent an unusual reservoir, and release the bleaching agent gradually into the bleaching bath. Moreover, the encapsulation of catalysts used for the decomposition of hydrogen peroxide radicals can be another factor in retarding the rate of decomposition. In this way, liposomes act as a stabilizing agent in the bleaching bath [34].

The role of auxiliary products is very important in textile dyeing with disperse dyes [53]. These compounds show extremely low solubility in water and dispersing agents are needed to maintain a fine, stable dispersion throughout the whole dyeing process at the different temperatures. Martí et al. (2007) analyzed the usefulness of commercial textile liposomes as dispersing agents, and observed that liposomes could be considered as suitable dispersing auxiliaries for polyester dyeing at high temperatures, considering their capacity to stabilize dye dispersions and achieve a suitable dye exhaustion level, with the added value of their environmentally friendly nature [53]. Liposomes clearly improve the dispersion efficiency as compared to conventional dispersing agents [34].

Additionally, liposomes for textile use show a similar price to that of synthetic surfactants used in the dyeing of polyester with disperse dyes. However, the new technology is more environmentally friendly, and hence the reduction in the environmental problem can lead to economic advantages [53]. In addition, liposome preparations tend not to foam. This is an advantage that distinguishes liposomes from other textile auxiliaries [34].

According to Martí et al. (2010), the dyeing of wool and wool blends with the aid of liposomes has demonstrated better quality, energy saving and a reduction in the environmental impact and also the temperature could be reduced, resulting in less fiber damage. Moreover, dye bath exhaustion was shown to be over 90% at the lower temperature (80°C) used, resulting in significant savings in energy costs [55]. The impact of the dyeing process on the environment was also considerably lower, the COD being reduced by about 1000 units [53].

Therefore, liposome-based technology is an alternative, eco-friendly method, which could reduce the environmental impact, offering technical and economic advantages for the textile industry.

2.2. Effect of ultrasonic energy on the dyeing process

Ultrasound-assisted textile dyeing was first reported by Sokolov and Tumansky in 1941[59]. The basic idea of this technology is that ultrasound can enhance mass transfer by reducing the stagnant cores in the yarns. The improvements observed are generally attributed to cavitation phenomena and to other resulting physical effects such as dye dispersion (breaking up of aggregates with high relative molecular mass), degassing (expulsion of dissolved or entrapped air from the fiber capillaries), strong agitation of the liquid (reduction in thickness of the fiber-liquid boundary layer), and swelling (enhancement of dye diffusion rate inside the fiber) [59,60].

According to Vankar & Shanker (2008), ultrasound allows for process acceleration, obtaining the same or better results than existing techniques, but under less extreme conditions, i.e., lower temperatures and lower concentrations of the chemicals used. Wet textile processes assisted by ultrasound are of great interest to the textile industry for this reason [61], and Khatri et al. (2011) showed that the dyeing of polyester fiber using ultrasonic energy resulted in an increased dye uptake and enhanced dyeing rate [35].

Due to the revolution in environmental protection, the use of ultrasonic energy as a renewable source of energy in textile dyeing has been increased, due to the variety of advantages associated with it. On the other hand, there is a growing demand for natural, eco-friendly dyeing for the health sensitive application to textile garments as an alternative to harmful synthetic dyes, which poses a need for suitable effective dyeing methodologies [62].

Ultrasonic energy can clean or homogenize materials, accelerating both physical and chemical reactions, and these qualities can be used to improve textile processing methods. Environmental concern has been focused on textile processing methods for quite some time, and the use of ultrasonic energy has been widely studied in terms of improving washing fastness. The textile dyeing industry has long been struggling to cope with high energy costs, rapid technological changes and the need for a faster delivery time, and the effective management of ultrasonic energy could reduce energy costs and improve productivity [35]. Ultrasonic waves are vibrations with frequencies above 17 kHz, out of the audible range for humans, requiring a medium with elastic properties for propagation. The formation and collapse of the bubbles formed by ultrasonic waves (known as cavitation) is generally considered to be responsible for most of the physical and chemical effects of ultrasound in solid/liquid or liquid/liquid systems [63]. Cavitation is the formation of gas-filled microbubbles or cavities in a liquid, their growth, and under proper conditions, their implosive collapse [59].

It has been reported that ultrasonic energy can be applied successfully to wet textile processes, for example laundering, desizing, scouring, bleaching, mercerization of cotton fabrics, enzymatic treatment, dyeing and leather processing, together with the decoloration/mineralization of textile dyes in waste water [60].

In addition, ultrasonic irradiation shows promise, and has the potential, for use in environmental remediation, due to the formation of highly concentrated oxidizing species such as hydroxyl radicals ($HO\bullet$), hydrogen radicals ($H\bullet$), hydroperoxyl radicals ($HO_2\bullet$) and H_2O_2, and localized high temperatures and pressures [59]. Therefore, the use of ultrasonic energy could indeed reduce the environmental impact caused by the textile industry.

3. FINISHING AND WASTE WATER

The contamination of natural waters has become one of the biggest problems in modern society, and the economical use of this natural resource in production processes has gained special attention, since in predictions for the coming years, the amount of water required per capita is of concern. This environmental problem is related not only to its waste through misuse, but also to the release of industrial and domestic effluents [64].

Of the industries with high-polluting power, the textile dyeing industry, responsible for dyeing various types of fiber, stands out. Independent of the characteristics of the dyes chosen, the final operation of all dyeing process involves washing in baths to remove excesses of the original or hydrolyzed dyes not fixed to the fiber in the previous steps [36]. In these baths, as previously mentioned, it is estimated that approximately 10-50% of the dyes used in the dyeing process are lost, and end up in the effluent [17,21,22], contaminating the environment with about one million tons of these compounds [65]. The dyes end up in the water bodies due mainly to the use of the activated sludge treatment in the effluent treatment plants, which has been shown to be ineffective in removing the toxicity and coloring of some types of dye [33,60,66,67]. Moreover, the reduction of azo dyes by sodium hydrosulfite and the successive chlorination steps with hypochlorous acid, can form 2-benzotriazoles fenilbenzotriazol (PBTA) derivatives and highly mutagenic aromatic amines, often more mutagenic than the original dye [68]. In an aquatic environment, this dye reduction can occur in two phases: 1) The application of reducing agents to the newly-dyed fibers to remove the excess unbound dye, which could lead to "bleeding" of the fabrics during washing, and 2) The use of reducing agents in the bleaching process, in order to make the effluent colorless and conform with the legislation. This reduced colorless effluent containing dyes is sent to the municipal sewage treatment plant, where they chlorinate the effluents before releasing them into water bodies where they may generate PBTAs. Several different PBTAs are already described in the literature, and their chemical structures vary depending on the dyes that originated them [63,69].

So the release of improperly treated textile effluents into the environment can become an important source of problems for human and environmental health. The major source of dye loss corresponds to the incomplete fixation of the dyes during the textile fiber dyeing step [36].

In addition to the problem caused by the loss of dye during the dyeing process, within the context of environmental pollution, the textile industry is also focused due to the large volumes of water used by its industrial park, consequently generating large volumes of effluent [64].

It has been calculated that approximately 200 liters of water are needed for each kilogram of cotton produced [70]. These effluents are complex mixtures of many pollutants, ranging from original colors lost during the dyeing process, to associated pesticides and heavy metals [71], and when not properly treated, can cause serious contamination of the water sources [64]. So the materials that end up in the water bodies are effluents containing a high organic load and biochemical oxygen demand, low dissolved oxygen concentrations, strong color and low biodegradability. In addition to visual pollution, the pollution of water bodies with these compounds causes changes in the biological cycles of the aquatic biota, particularly affecting the photosynthesis and oxygenation processes of the water body, for example by hindering the passage of sunlight through the water [72].

Moreover, studies have shown that some classes of dye, especially azo dyes and their by-products, may be carcinogenic and / or mutagenic [27,33,67,73-77], endangering human health, since the wastewater treatment systems and water treatment plants (WTP) are ineffective in removing the color and the mutagenic properties of some dyes [78,79]. The difficulty in removing them from the environment can be attributed to the high stability of these compounds, since they are designed to resist biodegradation to meet the demands of the consumer market with respect to durability of the colors in the fibers, consequently implying that they also remain in the environment for a long time [32].

With respect to the legislation, there is no consensus amongst the different countries concerning effluent discharge, and there is no official document listing the different effluent limit values applied in different countries. Many federal countries, such as the United States of America, Canada and Australia have national environmental legislation, which, as in Europe, establishes the limits that must be complied with. Some countries, such as Thailand, have copied the American system, whereas others, such as Turkey or Morocco, have copied the European model. In some countries, for example India, Pakistan and Malaysia, the emission limits are recommended, but are not mandatory [80]. With respect to the color, in some countries such as France, Austria and Italy, there are limits for the color of the effluent, but since they use different units, a comparison is impossible. The oldest unit is the Hazen, in use since the beginning of the 20th century, but in France, the current unit is (mg L^{-1}Pt–Co). The coloration values are determined by a comparative analysis with model solutions prepared according to defined procedures [80].

Based on all the problems cited above regarding the discharge of effluents into the environment, it is obvious there is a need to find alternative treatments that are effective in removing dyes from effluents.

4. WASTE WATER TREATMENT

These days, environmental pollution can undoubtedly be regarded as one of the main problems in developed and developing countries. This is due, not just to one, but to a number of factors, such as the misuse of natural resources, inefficient legislation and a lack of environmental awareness. Fortunately, in recent years there has been a trend for change and a series of scientific studies are being used as an important tool in the development of new treatment technologies and even in the implementation of processes and environmentally friendly actions [64,81-84].

Every industrial process is characterized by the use of inputs (raw materials, water, energy, etc.) that undergo transformation giving rise to products, byproducts and waste. The wastes produced at all stages of the various types of human activity, both in terms of composition and volume, vary according to the consumption practices and production methods. The main concerns are focused on the impact these can have on human health and the environment. Hazardous waste, produced mainly by industry, is particularly worrying, because when incorrectly managed, it becomes a serious threat to the environment and therefore to human health. Thus the study of new alternatives for the treatment of different types of industrial effluent continues to be a challenge to combat anthropogenic contamination.

Amongst that of several other industries, the textile sector waste has received considerable attention in recent years, since it can generate large volumes of effluents that, if not correctly treated before being disposed into water resources, can be a problem, as previously mentioned. Effluents from the textile industry are extremely complex, since they contain a large variety of dyes, additives and derivatives that change seasonally, increasing the challenge to find effective, feasible treatments. Currently, the processes developed and available for these industries are based on methods that were designed for other waste, and have limitations when applied to textile effluents. As a consequence, these industries produce colored wastewater with a high organic load, which can contribute enormously to the environmental pollution of surface water and treatment plants if not properly treated before disposal into the water resources [85]. The ingestion of water contaminated with textile dyes can cause serious damage to the health of humans and of other living organisms, due to the toxicity, highlighting mutagenicity of its components [86,87]. Therefore treatments that are more efficient and economical than those currently available are required.

There are several techniques for the treatment of effluents, such as incineration, biological treatment, absorption onto solid matrices, etc. However, these techniques have their drawbacks, such as the formation of dioxins and furans, caused by incomplete combustion during incineration; long periods for biological treatment to have an effect, as also the adsorptive process, that is based on the phase transfer of contaminants without actually destroying them [88,89]. The problem is further aggravated in the textile industry effluents, due to the complexity of their make-up. Thus it can be seen that processes are being used that are not entirely appropriate for the treatment of textile effluents, thereby creating a major challenge for the industry and laundries that need to adapt to current regulations for the control of the color of effluents with a high organic load.

The use of filtration membranes and/or separation [90] and biological methods [91], in addition to incineration processes involving adsorption onto solid matrices, has also being adopted by the textile industry and is receiving considerable attention. However, all these processes only involve phase transfer, generating large amounts of sludge deposited at the end of the tanks and low efficiency in color removal and reduction of the organic load. According to this scenario, many studies have been carried out with the aim of developing new technologies capable of minimizing the volume and toxicity of industrial effluents. Unfortunately, the applicability of these types of system is subject to the development of modified procedures and the establishment of effluent recycling systems, activities that imply evolutionary technologies and which are not yet universally available. Thus the study of new alternatives for the treatment of many industrial effluents currently produced is still one of the main weapons to combat the phenomenon of anthropogenic contamination.

Due to their considerable danger, several authors have attempted to find new forms of treatment to reduce the serious environmental and toxicological risks caused by various organic compounds. Amongst the many reported cases are those based on the use of specific microorganisms, and degradation using advanced oxidation processes (AOP) such as Fenton, photo-Fenton and heterogeneous photocatalysis, which are highlighted below.

The use of microorganisms cultivated specifically for the degradation of polluents to increase the yield of degradation, has been reported by some authors. For example, Flores et al. (1997) examined the behavior of 25 N-substituted aromatic compounds such as organic compounds,azo dyes and nitro, using the methanogenic bacteria acetoclastic, and found that under anaerobic conditions it was easy to mineralize various of the compounds evaluated with a good yield, especially the nitroaromatic and azo dyes [91].

Bornick et al. (2001) evaluated the use of aerobic microorganisms to degrade aromatic amines present in sediments of the river Elbe in Germany. The results obtained showed that it was possible to predict the qualitative degradation of the aromatic amines using degradation constants [92].

Using a structural design, Wang et al.(2007) isolated a bacterium capable of promoting the degradation of the compounds pentyl amine and aniline present in water oil extraction in China. Under conditions of neutral pH and complete aeration of 6 mg O_2 / l at a temperature of 30 ° C, they obtained degradation yields of 82% and 78%, respectively, for pentyl amine and aniline [93].

However, in general, although the use of microorganisms in the treatment of industrial and laboratory wastes containing aromatic amines deserves attention, mainly due to the low investment and maintenance costs, the results are far from ideal, due to the low biodegradation yields, long treatment times, and the generation of sludge deposited at the bottom of the treatment ponds [94].

Of the studies carried out using a source of hydroxyl radicals with oxidizing agents, the photo Fenton system developed by Fukushima et al., 2000 to promote the degradation of aniline stands out. This method has shown promise for the mineralization of aromatic amines, obtaining a reduction of approximately 85%. However, high performance liquid chromatography (HPLC) identified a number of intermediate species formed during the degradation of the aniline, such as p-aminophenol, p-hydroquinone, maleic and fumaric acids and NH_4^+ [95].

Studies involving heterogeneous photocatalysis also deserve attention: for example Pramauro et al., (1995) promoted the degradation of various aniline derivatives using TiO_2 particles suspended in a solution. Under optimal conditions, the method developed showed rapid mineralization of the aromatic amines examined in less than 1 hour of analysis. By-products generated during the degradation of these compounds were identified at the start of the reaction, but none were identified at the end of the reaction. The use of solar radiation was also evaluated, but was shown to be less efficient than artificial radiation [96].

Augugliaro et al.(2000), confirmed that heterogeneous photocatalysis using TiO_2 as a semiconductor may be a suitable method for the complete photodegradation of aniline, 4-ethylaniline and 4-chloroaniline in an aqueous medium. The kinetic parameters for the Langmuir-Hinshelwood model were used to describe the importance of the adsorption results, which proved to be independent of the pH of the solution and of the type of substituent on the aromatic ring of the amine [97].

Canle et al. (2005) studied the behavior of the adsorption of aniline and dimethyl aniline onto three species of TiO_2 (P25, anatase and rutile) and established that the greatest adsorption occurred onto TiO_2 P25. The effect of pH on the degradation provided by these compounds was also evaluated, and lower mineralization percentages were observed in acidic media. This type of behavior can be attributed to the positive charges on both the aniline and the semiconductor, providing electrostatic repulsion between the two species. Thus, an alkaline medium has been recommended as the most appropriate one to promote the mineralization of aromatic amines using a TiO_2 P25 type semiconductor [98].

Chu et al. (2007) observed the effects of pH variation and the addition of hydrogen peroxide on the degradation of 2-chloroaniline using TiO_2 as a semiconductor, with and without the application of UV radiation. The results showed that the addition of low concentrations of H_2O_2 to the UV/TiO_2system provided a significant increase in degradation of the aromatic amine. The addition of an excess of H_2O_2 promoted no increase in degradation, as expected; to the contrary, a reduction in the reaction rate was observed. The variation in pH was evaluated in both systems, and the condition leading to the highest percentages of mineralization was obtained in an alkaline medium using H_2O_2/UV/TiO_2 [99].

Low et al. (1991) monitored the inorganic products resulting from the degradation of several organic nitrogenated, sulfured and halogenated compounds. Degradation was carried out using TiO_2 as the semiconductor and artificially illuminated UV radiation. Ammonium ions were found to be present in higher concentrations than nitrate ions, which can be explained by the fact that compounds having a nitrogen element in their structure pass through a complex degradation step where the generation of ammonium ions is more favorable than the generation of nitrate ions. In turn, compounds with nitro groups in their structures, showed higher concentrations of nitrate ions. Under ideal conditions, all the elements were converted into their respective inorganic forms. Organic carbon was converted to CO_2, the halogens to their corresponding halide, sulfur compounds to sulfate, the phosphate to phosphorus and nitrogen to ammonium and nitrate [100]. However, all these studies used a photocatalytic titanium suspension, requiring a subsequent step to remove the semiconductor.

With a view to these problems, a new technique that has been studied recently, with significant success, is the oxidation of organic matter via the generation of hydroxyl radicals (OH•). This kind of effluent treatment has been highlighted for its destructive character with respect to the organic matter. The advanced oxidation processes (AOP) are processes based on the generation of highly oxidizing species, the hydroxyl radical (OH•), which can oxidize the organic contaminants present in water, air or soil. These radicals have a high oxidation potential (E * = +2.72 V vs. the normal hydrogen electrode, NHE), which results in high reactivity with organic pollutants. This may initiate different types of reaction with different functional groups in organic compounds, forming unstable organic radicals which are then easily oxidized to CO_2, H_2O and inorganic acids, derived from this heteroatom. Many techniques have being developed using this principle as the method of treatment, and have shown potential success in organic residue mineralization processes [92,96, 101-103].

The association of the electrochemical properties with the photocatalytic properties of a semiconductor, have allowed for the development of a promising system, representing a very efficient technique as an AOP method that has received much attention in recent years, assigned as a photoelectrochemical process. The process gained notoriety for the possibility of forming hydroxyl radicals via the oxidation of water. The technique is based on the action of ultraviolet light (hυ) on a semiconductor capable of generating charges and e^-/ H^+, whose separation is facilitated by applying a positive potential (E_{APP}), greater than the potential of a flat band photocatalytic material. The generation of a potential gradient in the photoactivated semiconductor directs the electrons to an auxiliary electrode (cathode), delaying recombination between the holes (h^+) generated in the valence band (VB), and making them available for oxidation processes and the generation of hydroxyl radicals of interest [104].

The results of this process were very promising because of the relatively short treatment time but with great efficiency, both in the removal of color and in the reduction of the organic load. However, the limitations of this technique are related mainly to the choice of the ideal catalyst for promoting the generation of these oxidizing species. Catalysts that promote the generation of radicals absorbing radiation in the visible spectral region are the most desirable for this type of reaction, due to the large percentage emitted in the solar spectrum (approximately 45%) [104].

Thus, the development of an ideal process that promotes color removal and a reduction in the organic load of wastewater from the textile industry with great efficiency is a major challenge in all fields of science, since the synthesis of the best catalyst to take advantage of solar radiation, thus reducing the operating costs, and at the same time solve the problems involved in the hydrodynamics of the reactors, is of importance in the development of the treatment.

The expectations for developing an effective method for the treatment of these wastes are quite promising, but require continuous optimization and knowledge of new aspects. These include better fixation of the dyes to the fibers, process with less water consumption, less hazardous dyes with respect to human health and methods capable of identifying these compounds with more efficacy and rapidity and assays to identify any potential carcinogenic and / or mutagenic properties in the dyes and their derivatives; genetic improvements to produce more efficient culture mediums and resistant biological treatments, leading to a reduction in the generation of sludge; the synthesis of materials that catalyze reactions in the visible spectral regions, leading to a more economic photoeletrochemical method, and also new engineering advances for the construction of more effective reactors, which can take advantage of all these developments in an integrated system, extending the performance of a process more appropriate for the treatment of such a complex effluent.

5. OPTIMIZATION OF THE DYEING PROCESSES TO REDUCE THE ENVIRONMENTAL IMPACT OF THE TEXTILE INDUSTRY

The search and development of new methods to promote the treatment of effluents from the textile industry with a maximum of efficiency of the process of decolorization and / or removal of these compounds present in the medium can trigger further damage human health and the environment is fundamental importance. The understanding of the composition of waste generated is extremely significant to develop these methods of treatment due to the high complexity by virtue of huge number of compounds which are added at different stages of the dyeing fabrics.

Environmental problems with used dye baths are related to the wide variety of different components added to the dye bath, often in relatively high concentrations. In the future, many of textile factories will face the requirement of reusing a significant part of all incoming freshwater because traditionally used methods are insufficient for obtaining the required water quality.

However, due to dwindling supply and increasing demand of water in the textile industries, a better alternative is to attempt to further elevate the water quality of wastewater effluent from a secondary wastewater treatment plant to a higher standard for reuse. Thus far very little attention has been paid to this aspect [105].

Therefore, the investment in the search for methodologies to more effective treatment of these effluents can be much smaller than that spent in tertiary treatment to remove these products in low level of concentrations and in the presence of much other interference. This requires action that the cost / benefit are reviewed and the development of new techniques for wastewater treatment capable of effective removal of these dyes is intensified and made economically viable [105,106].

An alternative to minimize the problems related to the treatment of textile effluents would be the development of more effective dye that can be fixed fiber with higher efficiency decreasing losses on tailings waters and reducing the amount of dye required in the dyeing process, reducing certainly improve the cost and quality of the effluent.

6. CONCLUSION

It was concluded that the synthetic textile dyes represent a large group of organic compounds that could have undesirable effects on the environment, and in addition, some of them can pose risks to humans. The increasing complexity and difficulty in treating textile wastes has led to a constant search for new methods that are effective and economically viable. However, up to the present moment, no efficient method capable of removing both the color and the toxic properties of the dyes released into the environment has been found.

ACKNOWLEDGMENTS

This work was supported by the Faculty of Pharmaceutical Sciences at Ribeirão Preto - University of São Paulo, Brazil, and by FAPESP, CAPES and CNPq.

REFERENCES

1. Kirk-Othmer. Encyclopedia of Chemical Technology, v. 7, 5th Edition. Wiley-Interscience; 2004.

2. Bafana A, Devi SS, Chakrabarti T. Azo dyes: past, present and the future. Environmental Reviews 2011; 19 350–370.

3. Zollinger, H. Synthesis, Properties of Organic Dyes and Pigments. In: Color Chemistry. New York, USA: VCH Publishers; 1987. p. 92-102.

4. Carneiro PA, Nogueira, RFP, Zanoni, MVB. Homogeneous photodegradation of C.I. Reactive Blue 4 using a photo-Fenton process under artificial and solar irradiation. Dyes and Pigments 2007; 74 127-132.

5. Christie R. Colour Chemistry. Cambridge, United Kingdom: The Royal Society of Chemistry; 2001.

6. Dos Santos AB, Cervantes FJ, van Lier JB. Review paper on current technologies for decolourisation of textile wastewaters: Perspectives for anaerobic biotechnology. Bioresource Technology 2007; 98 (12) 2369-2385.

7. Arun Prasad AS, Bhaskara Rao KV. Physico chemical characterization of textile effluent and screening for dye decolorizing bactéria. Global Journal of Biotechnology and Biochemistry 2010; 5(2) 80-86.

8. Robinson T, McMullan G, Marchant R, Nigam P. Remediation of dyes in textile effluent: a critical review on current treatment technologies with a proposed alternative. Bioresource Technology 2001; 77 (12) 247-255.

9. Ogugbue CJ, Sawidis T. Bioremediation and Detoxification of Synthetic Wastewater Containing Triarylmethane Dyes by Aeromonas hydrophila Isolated from Industrial Effluent. Biotechnology Research International 2011; DOI 10.4061/2011/967925.

10. Perkins WS. A Review of Textile Dyeing Processes. American Association of Textile Chemists and Colorists 1991; 23 (8): 23–27.http://www.revistavirtualpro.com/files/TIE08_200704.pdf (accessed 12 May 2012)

11. Couto SR. Dye removal by immobilised fungi. Biotechnology Advances 2009; 27(3) 227-235.

12. O'Neill C, Hawkes FR, Hawkes DL, Lourenço ND, Pinheiro HM, Delée W. Colour in textile effluents – sources, measurement, discharge consents and simulation: a review. Journal of Chemical Technology and Biotechnology 1999; 74 (11) 1009-1018.

13. Forgacs E, Cserháti T, Oros G. Removal of synthetic dyes from wastewaters: a review. Environment International 2004; 30 (7) 953- 971

14. Przystaś W, Zabłocka-Godlewska E, Grabińska-Sota E. Biological Removal of Azo and Triphenylmethane Dyes and Toxicity of Process By-Products. Water Air Soil Pollut 2012; 223 (4) 1581-1592.

15. Sen S, Demirer GN. Anaerobic treatment of real textile wastewater with a fluidized bed reactor. Water Research 2003; 37 (8) 1868-1878.

16. Dos Santos, A B. Reductive Decolourisation of Dyes by Thermophilic Anaerobic Granular Sludge. PhD-Thesis. Wageningen University, The Netherlands, 2005.

17. Ben Mansour H, Houas I, Montassar F, Ghedira K, Barillier D, Mosrati R, Chekir-Ghedira L. Alteration of in vitro and acute in vivo toxicity of textile dyeing wastewater after chemical and biological remediation. Environmental science and pollution research international 2012; DOI 10.1007/s11356-012-0802-7.

18. Talarposhti AM, Donnelly T, Anderson GK. Colour removal from a simulated dye wastewater using a two-phase anaerobic packed bed reactor. Water Research 2001; 35 (2) 425-432.

19. Ibrahim MB, Poonam N, Datel S, Roger M. Microbial decolorization of textile dye-containing effluents: a review, Bioresource Technology 1996; 58(3) 217-227.

20. Wijetunga S, Li XF, Jian C. Effect of organic load on decolourization of textile wastewater containing acid dyes in upflow anaerobic sludge blanket reactor. Journal of Hazardous Materials 2010; 177 (1-3) 792-798.

21. Vaidya AA, Datye KV. Environmental pollution during chemical processing of synthetic fibers. Colourage 1982; 14 3-10.

22. Rajaguru P, Fairbairn LJ, Ashby J, Willington MA, Turner S, Woolford LA, Chinnasamy N, Rafferty JA. Genotoxicity studies on the azo dye Direct Red 2 using the in vivo mouse bone marrow micronucleus test. Mutation Research 1999; 444(1) 175-180.

23. Hubbe MA, Beck KR, O'Neal WG, Sharma YC. Cellulosic substrates for removal of pollutants from aqueous systems: a review. 2. Dyes. Dye biosorption: Review. BioResources 2012; 7 (2), 2592-2687.

24. Carliell CM, Barclay SJ, Shaw C, Wheatley AD, Buckley C A. The effect of salts used in textile dyeing on microbial decolourisation of a reactive azo dye. Environmental Technology 1998; 19 (11) 1133-1137.

25. Seesuriyachan P, Takenaka S, Kuntiya A, Klayraung S, Murakami S, Aoki K. Metabolism of azo dyes by Lactobacillus casei TISTR 1500 and effects of various factors on decolorization. Water Research 2007; 41(5) 985-992.

26. Ben Mansour H, Corroler D, Barillier D, Ghedira K, Chekir L, Mosrati R. Evaluation of genotoxicity and pro-oxidant effect of the azo dyes: Acids yellow 17, violet 7 and orange 52, and of their degradation products by Pseudomonas putida mt-2. Food and Chemical Toxicology 2007; 45 (9) 1670-1677.

27. Chung KT, Cerniglia CE. Mutagenicity of azo dyes: Structure-activity relationships. Mutation Research, 1992; 277 (3) 201-220.

28. Pinheiro HM, Touraud E, Thomas O. Aromatic amines from azo dye reduction: status review with emphasis on direct UV spectrophotometric detection in textile industry wastewaters. Dyes and Pigments 2004; 61 (2) 121-139.

29. Umbuzeiro GA, Freeman H, Warren SH, Kummrow F, Claxton LD. Mutagenicity evaluation of the commercial product CI Disperse Blue 291 using different protocols of the Salmonella assay. Food and Chemical Toxicology 2005; 43 (1) 49-56.

30. Arlt VM, Glatt H, Muckel E, Pabel U, Sorg BL, Schmeiser HH, Phillips DH. Metabolic activation of the environmental contaminant 3-nitrobenzanthrone by human acetyltransferases and sulfotransferase. Carcinogenesis 2002; 23 (11) 1937-1945.

31. Hao OJ, Kim H, Chiang PC, 2000. Decolorization of wastewater. Critical Reviews in Environmental Science and Technology 2000; 30 (4) 449-505.

32. Firmino PIM, Silva MER, Cervantes FJ, Santos AB. Colour removal of dyes from synthetic and real textile wastewaters in one- and two-

stage anaerobic systems. Bioresource Technology 2010; 101(20) 7773-7779.

33. Carneiro PA, Umbuzeiro GA, Oliveira DP, Zanoni MVB. Assessment of water contamination caused by a mutagenic textile effluent/dyehouse effluent bearing disperse dyes. Journal of Hazardous Materials 2010; 174 (1-3) 694-699.

34. Barani H, Montazer M. A Review on Applications of Liposomes in Textile Processing. Journal of Liposome Research 2008; 18 (3) 249-262.

35. Khatri Z, Memon MH, Khatri A, Tanwari A. Cold Pad-Batch dyeing method for cotton fabric dyeing with reactive dyes using ultrasonic energy. Ultrasonics Sonochemistry 2011 18 (6) 1301-1307.

36. Guaratini CCI, Zanoni MVB. Textile dyes. Química Nova 2000;23(1): 71-78. http://www.scielo.br/pdf/qn/v23n1/2146.pdf (accessed 01 May 2012).

37. Moore SB, Ausley LW. Systems thinking and green chemistry in the textile industry: concepts, technologies and benefits. Journal of Cleaner Production 2004;12 585–601. DOI: 10.1016/S0959-6526(03)00058-1

38. Reddy SS, Kotaiah B, Reddy NSP. Color pollution control in textile dyeing industry effluents using tannery sludge derived activated carbon. Bulletin of the Chemical Society of Ethiopia 2008;22 (3): 369-378. http://www.ajol.info/index.php/bcse/article/ viewFile/ 61211/49389 (accessed 16 May 2012).

39. Alcantara MR, Daltin D. A química do processamento têxtil. Química Nova 1996;19(3): 320-330 http://www.quimicanova.sbq. org.br/ qn/qnol/1996/vol19n3/v19_n3_17.pdf (accessed 19 April 2012).

40. Vassileva V, Valcheva E, Zheleva Z. The kinetic model of reactive dye fixation on cotton fibers. Journal of the University of Chemical Technology and Metallurgy 2008;43 (3): 323-326. http:// www.uctm.edu/j2008-3/8_V_Vasileva_323-326.pdf (accessed 10 May 2012).

41. Zollinger H. Color Chemistry. Syntheses,properties and applications of organic dyes and pigments. Weinheim: VCH Publishers; 2003.

42. Candlin J. Polymers. Dyestuffs in the Chemical Industry. London: Chapman & Hall; 1994.

43. Gregory P. Dyestuffs. In: Heaton C.A. (ed.) The Chemical Industry. London: Chapman & Hall; 1994. p143-188.

44. Gregory P. 2009. Dyes and Dye Intermediates. In: Kirk-Othmer (ed.) Encyclopedia of Chemical Technology. New Jersey: John Wiley & Sons; 2009. p1–66.

45. Pang K, Kotek R, Tonelli A. Review of conventional and novel polymerization processes for polyesters. Progress in Polymer Science 2006;31(11) 1009–1037.

46. Hunger K., editor. Industrial Dyes: Chemistry, Properties, Applications. Weinheim: Wiley-VCH Verlag GmbH & Co. KGaA; 2004.

47. Baptista ALF, Coutinho PJG,: Real Oliveira MECD, Rocha Gomes JIN. Effect of surfactants in soybean Lecithin liposomes studied by energy transfer between NBD-PE and N-Rh-PE. Journal of Liposome Research 2000; 10 (4) 419-429.

48. Lesoin L, Crampon C, Boutin O, Badens E. Preparation of liposomes using the supercritical anti-solvent (SAS) process and comparison with a conventional method. The Journal of Supercritical Fluids 2011; 57 162-174

49. Sivasankar, M, Katyayani, T. Liposomes – The future of formulations. International Journal of Research in Pharmacy and Chemistry 2011; 1(2) 259-267.

50. El-Zawahry MM, El-Shami S, El-Mallah S.H. Optimizing a wool dyeing process with reactive dye by liposome microencapsulation. Dyes and Pigments 2007; 74 684-691.

51. Martí M, de la Maza A, Parra JL, Coderch L. Liposome as dispersing agent into disperse dye formulation. Textile Research Journal 2011; 81(4) 379-387.

52. Montazer M., Validi M., Toliyat T. Influence of Temperature on Stability of Multilamellar Liposomes in Wool Dyeing. Journal of Liposome Research, 2006; 16:81-89.

53. Martí M, Coderch L, de la Maza A, Parra JL. Liposomes of phosphatidylcholine: a biological natural surfactant as a dispersing agent. Color Technology 2007; 123 237-241.

54. Riaz M. Liposomes preparation methods. Pakistan Journal of Pharmaceutical Sciences, 1996; 19(1) 65-77.

55. Martí M, de la Maza A, Parra JL, Coderch L. Liposome as dispersing agent into disperse dye formulation. Textile Research Journal 2010; 81(4) 379–387.

56. Nelson, G. Application of microencapsulation in textile. International Journal of Pharmaceutics 2002; 242 55-62.

57. Martí M, de la Maza A, Parra JL, Coderch L. Dyeing wool at low temperatures: new method using liposomes. Textile Research Journal 2001; 71(8) 678-682.

58. El-Zawahry MM, El-Mallahb MH, El-Shamib S. An innovative study on dyeing silk fabrics by modified phospholipid liposomes. Coloration Technology 2009; 125 164-171.

59. Vajnhandl S, Le Marechal AM. Ultrasound in textile dyeing and the decolouration/mineralization of textile dyes. Dyes and Pigments 2005; 65 89-101.

60. Ferrero F., Periolatto M. Ultrasound for low temperature dyeing of wool with acid dye. Ultrasonics Sonochemistry 2012; 19 601-606.

61. Vankar PS, Shanker R. Ecofriendly ultrasonic natural dyeing of cotton fabric with enzyme pretreatments. Desalination 2008; 230 62-69.

62. Mansour HF, Heffernan S. Environmental aspects on dyeing silk fabric with sticta coronate lichen using ultrasonic energy and mild mordants. Clean Technologies Environmental Policy 2011; 13 207-213.

63. Abou-Okeil A, El-Shafie A, El Zawahry MM. Ecofriendly laccase–hydrogen peroxide/ultrasound-assisted bleaching of linen fabrics and its influence on dyeing efficiency. Ultrasonics Sonochemistry 2010; 17 383-390.

64. Kunz A, Zamora PP, Moraes SG, Durán N. New tendencies on the textile effluent treatment. Quimica Nova 2002;25(1): 78-82. http://www.scielo.br/scielo.php?script=sci_arttext&pid=S0100-40422002000100014 (accessed 01 May 2012).

65. Associação Brasileira de Química. ABQ. Processos oxidativos avançados e tratamento de corantes sintéticos. Anais da ABQ. http://www.unb.br/resqui/abq2004-1.pdf (accessed 18 May 2012).

66. Umbuzeiro GA, Roubicek DA, Rech CM, Sato MIZ, CLAXTON LD. Investigating the sources of the mutagenic activity found in a river using the Salmonella assay and different water extraction procedures. Chemosphere 2004; 54(11) 1589-1597.

67. Umbuzeiro GA, Freeman HS, Warren SH, Oliveira DP, Terao Y, Watanabe T, Claxton LD. The contribution of azo dyes to the mutagenic activity of the Cristais river. Chemosphere 2005; 60 (1) 55-64.

68. Shiozawa T, Suyama K, Nakano K, Nukaya H, Sawanishi H, Oguri A, Wakabayashi K, Terao Y. Mutagenic activity of 2 phenylbenzotriazole derivatives related to a mutagen, PBTA-1, in river water. Mutation Research 1999; 442 (2) 105-111.

69. Oliveira DP. Corantes como importante classe de contaminantes ambientais – um estudo de caso. PhD thesis. Universidade de São Paulo; 2005.

70. Carneiro PA, Osugi ME, Sene JJ, Anderson MA, Zanoni MVB. Evaluation of color removal and degradation of a reactive textile azo dye on nanoporous TiO thin-film electrodes. Electrochimica Acta 2004; 49 (22-23) 3807–3820.

71. McMullan G, Meehan C, Conneely A, Kirby N, Robinson T, Nigam P, Banat IM, Marchant R, Smyth WF. Mini-review: microbial decolorisation and degradation of textile dyes. Applied Microbiology and Biotechnology 2001; 56 (1-2): 81-87. http://www.springerlink.com/content/04vuqljhajpa32rv/ (accessed 10 April 2012).

72. Pereira WS, Freire RS. Ferro zero: Uma nova abordagem para o tratamento de águas contaminadas com compostos orgânicos poluentes. Química Nova 2005;28(1): 130-136. http:// www.

scielo.br/scielo.php?script=sci_arttext&pid=S0100-40422005000100022 (accessed 10 March 2012).

73. Houk VS. The genotoxicity of industrial wastes and effluents. Mutation Research 1992; 277 (2) 91-138.

74. Jager I, Hafner C, Schneider K. Mutagenicity of different textile dye products in Salmonella typhimurium and mouse lymphoma cells. Mutation Research 2004;561 (1-2) 35-44.

75. Chequer FMD, Angeli JPF, Ferraz ERA, Tsuboy MS, Marcarini JC, Mantovani MS, Oliveira DP. The azo dyes Disperse Red 1 and Disperse Orange 1 increase the micronuclei frequencies in human lymphocytes and in HepG2 cells. Mutation Research 2009; 676 (1-2) 83-86.

76. Chequer FMD, Lizier TM, Felicio R, Zanoni MVB, Debonsi HM, Lopes NP, Marcos R, Oliveira DP. Analyses of the genotoxic and mutagenic potential of the products formed after the biotransformation of the azo dye Disperse Red 1. Toxicology in Vitro 2011; 25 (8) 2054-2063

77. Ferraz ERA, Umbuzeiro GA, de-Almeida G, Caloto-Oliveira A, Chequer FMD, Zanoni MVB, Dorta DJ, Oliveira DP. Differential Toxicity of Disperse Red 1 and Disperse Red 13 in the Ames Test, HepG2 Cytotoxicity Assay, and Daphnia Acute Toxicity Test. Environmental Toxicology 2011; 26 (5) 489-497

78. Oliveira GAR, Ferraz ERA, Chequer FMD, Grando MD, Angeli JPF, Tsuboy MS, Marcarini JC, Mantovani MS, Osugi ME, Lizier TM, Zanoni, MVB, Oliveira DP. Chlorination treatment of aqueous samples reduces, but does not eliminate, the mutagenic effect of the azo dyes Disperse Red 1, Disperse Red 13 and Disperse Orange 1. Mutation Research 2010 703 (2) 200-208

79. Konstantinou IK, Albanis TA. TiO2-assisted photocatalytic degradation of azo dyes in aqueous solution: kinetic and mechanistic investigations - A review. Applied Catalysis B: Environmental 2004; 49 (1) 1-14.

80. Hessel C, Allegre C, Maisseu M, Charbit F, Moulin P. Guidelines and legislation for dye house effluents. Journal of Environmental Management 2007; 83 (2) 171-180.

81. Carneiro PA, Osugi ME, Fugivara CS, Boralle N, Furlan M, Zanoni MVB. Evaluation of different electrochemical methods on the oxidation and degradation of Reactive Blue 4 in aqueous solution. Chemosphere. 2005; 59(3) 431-439.

82. Mohan SV, Bhaskar YV, Karthikenyan J. Biological decolourization of simulated azo dye in aqueous phase by algae Spirogyra species. International Journal of Environment and Pollution. 2004; 21 (3) 211-222.

83. Vaghela SS, Jethva AD, Mehta BB, Dave SP, Adimurthy S, Ramachandraiah G. Laboratory studies of electrochemical treatment of industry azo dye effluent. Environmental Science and Technology. 2005; 39 (8) 2848-2855.

84. Cardoso JC, Lizier TM, Zanoni MVB. Highly ordered TiO2 nanotube arrays and photoelectrocatalytic oxidation of aromatic amine. Applied Catalysis, B: Environmental. Applied Catalysis, B: Environmental. 2010; 99 (1-2) 96-102.

85. Pereira RS. Identificação e caracterização das fontes de poluição em sistemas hídricos. Revista Eletrônica de Recursos Hídricos. IPH-UFRGS 2004; 1(1): 20-36. http://www.abrh.org.br/ informacoes/ rerh.pdf (accessed 06 June 2012).

86. Baumgarten, MGZ.; Pozza, SA. Qualidade de Águas. Descrição de Parâmetros Químicos referidos na Legislação Ambiental. Ed: FURG; 2001.

87. Prado AGS, Torres JD, Faria EA, Dias SCL. Comparative adsorption studies of indigo carmine dye on chitin and chitosan. Journal of Colloid and Interface Science. 2004; 1(1) 43-47.

88. Babel S, Kurniawan TA. Low-cost adsorbents for heavy metals uptake from contaminated water: a review. Journal of Hazardous Materials. 2003; 97(1-3) 219-243.

89. Associação Brasileira da Indústria Têxtil e de Confecção. Economia. <http://www.abit.org.br/site/navegacao.asp?id_menu=8&IDIOMA= PT> (accessed 19 August 2011).

90. Walsh FC. Electrochemical technology for environmental treatment and clean energy conversion. Pure and Applied Chemistry. 2001; 73 (12) 1819-1837.

91. Flores ER, Perez F, Torre M. Scale-up of Bacillus thuringiensis fermentation based on oxygen transfer. Journal of Fermentation and Bioengineering. 1997; 83 (6) 561-564.

92. Bornick H. Simulation of biological degradation of aromatic amines in river bed sediments. Water Research. 2001; 35 (3) 619-624.

93. Wang L, Barrington S, Kim JW. Biodegradation of pentylamine and aniline from petrochemical wastewater. Journal of Environmental Management. 2007; 83 (2) 191-197.

94. Orshansky F, Narkis N. Characteristics of organics removal by pact simultaneous adsorption and biodegradation. Water Research. 1997; 31(3) 391-398.

95. Fukushima M, Tatsumi K, Morimoto K. The fate of aniline after a photo-fenton reaction in an aqueous system containing iron(III), humic acid, and hydrogen peroxide. Environmental Science and Technology. 2010; 34 (10) 2006-2013.

96. Pramauro E. et al. Photocatalytic treatment of laboratory wastes containing aromatic amines. Analyst. 1995; 120 (2) 237-242.

97. Augugliaro V. et al. Photodegradation kinetics of aniline, 4-ethylaniline, and 4-chloroaniline in aqueous suspension of polycrystalline titanium dioxide. Research on Chemical Intermediates. 2000; 26 (5) 413-426.

98. Canle LM, Santaballa JA, Vulliet E. On the mechanism of TiO2-photocatalyzed degradation of aniline derivatives. Journal of Photochemistry and Photobiology, A: Chemistry. 2005; 175 (2-3) 192-200.

99. Chu W, Choy WK, So TY. The effect of solution pH and peroxide in the TiO2-induced photocatalysis of chlorinated aniline. Journal of Hazardous Materials. 2007; 141 (1) 86-91.

100. Low, G. K. C.; McEvoy, S. R.; Matthews, R. W. Formation of nitrate and ammonium ions in titanium dioxide mediated photocatalytic degradation of organic compounds containing nitrogen atoms. Environmental Science and Technology, v. 25, n. 3, p. 460-467, 1991.

101. Pinheiro HM, Touraud E, Thomas O. Aromatic amines from azo dye reduction: status review with emphasis on direct UV

spectrophotometric detection in textile industry wastewaters. Dyes and Pigments. 2004; 61(2) 121-139.

102. Fukushima M, Tatsumi K, Morimoto K. The Fate of Aniline after a Photo-Fenton Reaction in an Aqueous System Containing Iron(III), Humic Acid, and Hydrogen Peroxide. Environmental Science and Technology. 2006; 34 (10) 2006-2013.

103. Chu W, Choy WK, So TY. The effect of solution pH and peroxide in the TiO2-induced photocatalysis of chlorinated aniline. Journal of Hazardous Materials. 2007; 141 (1) 86-91.

104. Finklea HO. Semiconductor electrodes. New York: Elsevier; 1998.

105. Lin SH, Chen ML. Purification of textile wastewater effluents by a combined Fenton process and the ion exchange. Desalination 1997; 109 (1) 121-130.

106. Mozia S. Photocatalytic membrane reactor (PMRs) in water and wastewater treatment. A review. Separation and Purification Technology 2010; 73 (2) 71-91.

CHAPTER 2

Multifunctional Textiles – Modification by Plasma, Dyeing and Nanoparticles

Marija Gorjanc[1], Marija Gorenšek[1], Petar Jovančić[2] and Miran Mozetič[3]

[1] University of Ljubljana, Faculty of Natural Sciences and Engineering, Department of Textiles, Ljubljana, Slovenia

[2] University of Belgrade, Faculty of Technology and Metallurgy, Belgrade, Serbia

[3] Jozef Stefan" Institute, Ljubljana, Slovenia

1. INTRODUCTION

The textile industry in developed countries is confronting the world's marketing conditions and competitive challenges which are driving towards the development of advanced, highly functional textiles and textiles with higher added value. The conventional textile finishing techniques are wet chemical modifications where water and rather hazardous chemicals are used in large quantities and wastewaters need to be processed before discharging effluent, whereas the most problematic factor are ecological impacts to the environment and effects to human health. The increasing environmental concerns and demands for an environmentally friendly processing of textiles leads to the development of new technologies, the use of plasma being one of the suitable methods [1]. Plasma technology is an environmentally friendly technology and a step towards creating solid surfaces with new and improved properties that cannot be achieved by conventional processes [2]. Plasma is the fourth state of matter. It is a gas with a certain portion of ionized as well as other reactive particles, e.g. ions, electrons, photons, radicals and metastable excited particles. Several types of plasma are known; however, only non-equilibrium or cold plasma is used for the modification of physical and chemical properties of solid materials such as textiles. Chemically reactive particles produced at a low gas temperature are a unique property of cold plasma; hence, there is minimal thermal degradation of a textile substrate during the plasma processing [3]. Cold plasma is a partially ionized gas with the main characteristic of a very high temperature of free electrons (typically of the order of 10,000 K, often about 50,000 K) and a low kinetic temperature of all other species. The average energy of the excited molecules is usually far from the values calculated from the thermal equilibrium at room temperature. The rotational temperature, for instance, is often close to 1000 K, while the vibrational temperature can be as high as 10,000 K, although the kinetic (translation) temperature is close to room temperature. Furthermore, the dissociation fraction is often several percent, which is orders of magnitude larger than that calculated from thermal equilibrium at room temperature. This also applies to the ionization fraction; although this is often much lower than the dissociation fraction. Plasma with such characteristics readily interacts with solid surfaces, causing reactions that would otherwise occur only at elevated temperature of the solid material. For this reason, non-equilibrium plasma represents an extremely powerful medium for modification of the surface properties of solid materials. A medium of particular interest is weakly ionized highly dissociated oxidative plasma that can be sustained in high frequency discharges in oxygen, air, carbon dioxide, water vapor, and mixtures of these gases with a noble gas. Such plasma has been successfully utilized in an extremely wide range of applications from nanoscience to fusion reactors. Plasma is used for synthesis of nanostructures with interesting properties, removal of thin films of organic impurities, selective etching of composites, sterilization, passivation of metal, ashing of biological materials, etching of photoresists, functionalization of polymers, and conditioning of tokamaks with carbon walls [4-9].

The choice of discharge parameters is determined by the requirements of each particular application. For selective plasma etching, for instance, extremely aggressive plasma is needed; thus, it is created with powerful generators at a moderate pressure (where the O density is the highest) in pure oxygen or in a mixture of oxygen and argon. For treatment of delicate organic materials, on the other hand, weak plasma performs better, since aggressive plasma would destroy organic material in a fraction of a second. Therefore, extremely delicate organic materials are rather treated in an afterglow or in plasma created at low pressure and with a low-power generator. Water vapor is sometimes used instead of oxygen. The advantages of using plasma are ecological and economical. Moreover, the textiles subjected to the treatment are modified without an alteration of the bulk properties. Unlike wet chemical processes, which penetrate deep into the fibers, plasma produces no more than a surface reaction, the properties given to the material being limited to the surface layer of a few nanometers [10]. The modification of textile substrates using plasma enables different effects on the textile surfaces from the surface activation to a thin film deposition via plasma polymerization. In the first stage of the treatment, plasma reacts with the substrate surface where active species and new functional groups are created, which can completely change the reactivity of the substrate [11]. The changes in the surface morphology of fibers can be induced by plasma etching process where the nano- or micro-roughness of fibers is formed [12]. The nanostructured textile surfaces have a higher specific surface area, which leads to new or improved properties of the treated surface, i.e. increased surface activity, hydrophilic or hydrophobic properties, and increased absorption capacity towards different materials, i.e. nanoparticles and nano-composites [13- 18].

When researching the deposition of nanoparticles onto different substrates, it is important to understand the adhesion of particles, which is dependent on the interaction mechanism with a material. The mechanism of nanoparticle adhesion has not been completely explained yet, since there are many different opinions among the theorists on the subject. Thus, it is generally considered that attractive forces and chemical bonds play an important role in the adhesion of particles [19]. The physical or mechanical adhesion of nanoparticles mostly occurs due to van der Walls or electrostatic forces, while the chemical adhesion of particles is a consequence of ionic, covalent, metallic and hydrogen bonds [20]. Moreover, the nanoparticles can penetrate into certain parts of the substrate, such as pores, holes and crevices, and they lock mechanically to the substrate. This adhesion mechanism, which is called mechanical interlocking, has been solved from the perspective of surface roughness effects [21]. Since plasma causes etching of the fibers and leads to an increase of the surface roughness higher adhesion properties towards metal or ceramic particles onto substrates can be achieved [22-30]. The adhesion of TiN (titanium nitride) onto PP (polypropylene) and PC (polycarbonate) was increased after a modification of substrates with argon low-pressure plasma [31]. The roughness after plasma treatment increased from 15 nm to 17 nm for PP and from 12 nm to 30 nm for PC. Consequently, contact angles decreased from 95° to 59° for PC and from 87° to 35° for PP. The surface modification of polyethylene terephthalate (PET) polymer was created by oxygen and nitrogen plasma at different treatment times [32]. The surface of PET polymer was modified in order to achieve improved attachment of fucoidan, which is a bioactive coating with antithrombogenic properties. The attachment of fucoidan was improved by oxygen plasma treatment, especially due to the surface roughening. The adhesion work, the surface energy and the surface polarity of PA6 (polyamide-6) fibers were improved by dielectric barrier discharge (DBD) treatment in helium at atmospheric pressure. Furthermore, a new structure was observed at the nanoscale, with an increased roughness and a larger surface area, favoring the adsorption [33]. The self-cleaning and UV protective properties of PET fibers were drastically improved after a modification of PET fibers with oxygen plasma and loading of TiO_2 prepared by an aqueous sol-gel process [34]. Cotton also showed self-cleaning properties after RF plasma and TiO_2treatment [35]. TiO_2 on textile substrates is also used for a biomedical application to improve antimicrobial effectiveness of the fabric [36]. By using oxygen radiofrequency plasma at a higher power input, the roughness of fibers increased and likewise the adhesion of TiO_2 onto treated fabric. Treatment of PA and PET with corona plasma increased the adhesion of colloidal silver which affected the antifungal protection of the fabrics [37]. The quantity of silver on plasma-treated fabric was three times higher than on untreated fabric.

Preparation of metal nanoparticles also enables the development of new biocides. Due to their large surface area and ability to detain moisture, the textile materials are an excellent environment for a microorganism growth. Microorganisms can cause milder, aesthetic unpleasantness to serious health related problems. Textile materials with an antimicrobial effectiveness are used for medical, military and technical textiles, textiles for sports and leisure and bedding. At nanotechnology researches in textiles, different forms of silver were used, such as metal silver nanoparticles, silver chloride (AgCl) and composite particles of silver and titanium dioxide (Ag-TiO_2) [3, 24, 25, 27, 38-47]. In the case of antimicrobial efficiency, the surface coating of nanosilver on titanium dioxide maximizes the number of particles per unit area in comparison with the use of an equal mass fraction of pure silver [48,49]. Different methods have been used for the deposition and loading of silver nanoparticles onto synthetic and natural textiles, i.e. sonochemical coating, sol-gel process, dip-coating, pad-batch and exhaustion method, the use of nanoporous structure of cellulose fibers as a nanoreactor for *in situ* synthesis of nanoparticles and plasma sputtering process [3, 27, 48-54]. The possibility of loading nanoparticles using exhaustion method started recently [55]. The exhaustion method is the best process for uniform distribution of nanoparticles and is especially appropriate to be used when the simultaneous application of nanoparicles and dye onto fabric is performed, and dyed and antimicrobial effective fabric is achieved at the same time [3, 24, 25, 27]. Depending on the desirable functionality of a functionalized fabric, a treating bath may contain only dye, dye and silver nanoparticles or silver nanoparticles alone. An exhaustion method for loading of silver nanoparticles onto textile substrate was also used on silk fibers [54]. In that research, different concentrations of colloidal silver were used (10, 25, 50 and 100 ppm) and the effect of medium pH on the silver nanoparticles uptake on the fibers was studied. The antimicrobial effectiveness of functionalized fibers was better for samples with higher silver concentration and for samples treated in a medium with a lower pH. Also the use of salt (NaCl) improved a uniform distribution of silver particles on the fibers' surface which consequently improved antimicrobial effectiveness of fabrics. A difference between pad-dry-cure and exhaustion method on adhesion and antimicrobial activity of fabrics was performed with commercial silver nanoparticles and a reactive organic-inorganic binder [45]. Results revealed that using the same initial concentration of silver, the pad-dry-cure method resulted in a much lower quantity of adsorbed silver nanoparticles in comparison to the exhaust method. Another possible method for applying silver onto textiles is plasma polymerization method where surface of textile is functionalized with nanostructured silver film by magnetron sputtering [56].

When dealing with modification of textiles by silver it is important to know how the functionalization will affect the color change of fabric. A reflectance (UV/VIS) spectrophotometry is one of the methods to be used when detecting or controlling the presence of silver nanoparticles in a solution or on a textile substrate. It is a direct measure that the abundance of silver is in the topmost layer of the textile fabric [30]. The instrument analyzes the light being reflected from the sample and produces an absorption spectrum. Some of the electrons in the nanoparticles are not bound to the selected atom of silver, but are forming an electronic cloud. Light falling on these electrons excite the collective oscillations, called surface plasmons. The resonance condition is established when the frequency of light photons matches the natural frequency of surface electrons oscillating against the restoring force of positive nuclei. Surface plasmon resonance is the basis of many standard tools for measuring adsorption of material onto planar metal (typically gold and silver) surfaces or onto the surface of metal nanoparticles. It is the fundamental principle behind many color-based biosensor applications. As a result of the particles growth, an intense absorption band at 400 nm to 415 nm caused by collective excitation of all free electrons in the particles was observed [57]. Increase in diameters of the nanoparticles from 1 to 100 nm induces a shift of the surface plasmon absorption band to higher wavelength [50, 57-59]. That means that the size of nanoparticles is defined by their optical response, therefore by that the color that can be seen [50, 60]. Loading of colloidal silver nanoparticles onto bleached cotton fabric caused yellowish coloring of fabric with absorption maximum at 370 nm [24,30]. The evaluation of color changes of textiles modified by silver nanoparticles was also determined in CIELAB color space [60]. Loading of colloidal silver nanoparticles in concentration of 10 ppm induced very small, eye insensitive, color change ($\Delta E^* < 1$) to the fabric. When colloidal silver nanoparticles in concentration of 50 ppm were loaded, the color change of fabric was obvious ($\Delta E^* = 15.09$). Loading of silver onto textiles before or after dyeing also causes color changes; however the changes are not as extensive as with the white fabric. When colloidal silver was loaded before dyeing, the color change was $\Delta E^* = 1.44$ and when loaded after dyeing, the color change was $\Delta E^* = 2.73$. Color changes of textiles can be also induced by plasma treatment. When plasma is used for a modification of cotton to improve its hydrophilicity, then cotton has a better dyeability and consequently deeper coloration [61]. Also, a raw cotton fabric can be bleached by using ozone plasma [62]. The whiteness index (CIE WI) of fabric was even higher after plasma treatment than after peroxide bleaching (CIE WI O_3 = 95.3; CIE WI H_2O_2 = 94.5).

Since the modification of substrates by plasma improves the adhesion of metal nanoparticles onto substrates than it is no surprise that plasma modification of textiles has a special significance when applying silver nanoparticles. The chapter presents the influence of plasma treatment on loading capacity of cotton toward different forms of nanosilver. Exhaust dyeing process was used for loading of nanosilver onto plasma-treated cotton, which represents a new approach to textile finishing in achieving multifunctional textile properties.

2. EXPERIMENTAL SETUP AND METHODOLOGY

Low-pressure plasma of different working gases and air atmospheric corona plasma were used for a modification of textiles. Untreated and plasma-treated textile substrates were additionally modified by loading of silver nanoparticles during dyeing process. Morphological, chemical and physical properties of plasma-treated textile substrates were studied using microscopy (SEM), spectroscopy (XPS) and measuring of the breaking strength and elongation of textiles. The quantity of adsorbed silver was determined using mass spectrometry (ICP-MS), while the antibacterial efficiency of functionalized textiles was determined using microbiological tests.

2.1. Plasma Modification of Textiles

Cotton substrates were modified using different plasma systems, i.e. atmospheric air corona plasma [63-65] and low-pressure inductively coupled radiofrequency (ICRF) discharge plasma of different working gases, water vapor [3, 26, 66, 67] and tetrafluoromethane [27], respectively. ICRF plasma is particularly suitable for treatment of delicate materials with a large surface for the following reasons: the neutral gas kinetic temperature remains close to the room temperature; the plasma-to-floating potential difference is small; the density of neutral reactive particles is large; and extremely high treatment uniformity is achieved. The use of low-pressure plasma is a contemporary technological process, not yet fully applied in textile industry due to its discontinuous process. For a continuous, on-line processing interfaced to a conventional production line the use of corona atmospheric pressure plasma is recommended. Furthermore, the corona plasma treatment also introduces new functional groups onto fibers surfaces, produces surface cleaning and etching effect of treated textiles. The comparative research of corona plasma treatment of polyester was presented as well [24].

2.1.1. Low-Pressure Plasma Treatment

A RF generator with a nominal power of 5 kW and a frequency of 27.12 MHz was applied. The power absorbed by plasma, however, was much smaller due to poor matching and was estimated to about 500 W. The discharge chamber was a cylindrical Pyrex tube with a diameter of 27 cm and a length of 30 cm. Cotton fabric was put onto a glass holder mounted in the center of the discharge chamber. After closing the chamber the desired pressure of 0.4 mbar was achieved by a two-stage rotary pump with a nominal pumping speed of 65 m^3/h. The pressure was fairly stable during the experiment. Although the ultimate pressure of the rotary pump was below 1 Pa, the pressure remained much higher at 40 Pa. The source of water vapor was the cotton fabric itself. When tetrafluoromethane was used as a working gas, it was leaked into the chamber in order to obtain a pressure of about 100 Pa, the pressure where plasma is most reactive. By switching on the RF generator, the gas in the discharge chamber was partially ionized and dissociated, starting the plasma treatment of the fabric. The plasma treatment time was 10 s in both cases. The schematic diagram of low-pressure RF plasma reactor is presented in Fig. 1.

2.1.2. Atmospheric Plasma Treatment

Textile samples were treated in a commercial device, Corona-Plus CP-Lab MKII (Vetaphone, Denmark) (Fig. 2). Samples (270×500 mm^2) were placed on a backing roller (the electrode roll covered with silicon coating), rotating at a working speed of 4 m/min. The distance between electrodes was adjusted with air gap adjusters at both sides of the electrode to 2 mm. Corona discharge was generated within the air gap between the electrode and backing roller. The power was 900 W and the number of passages was set to 30.

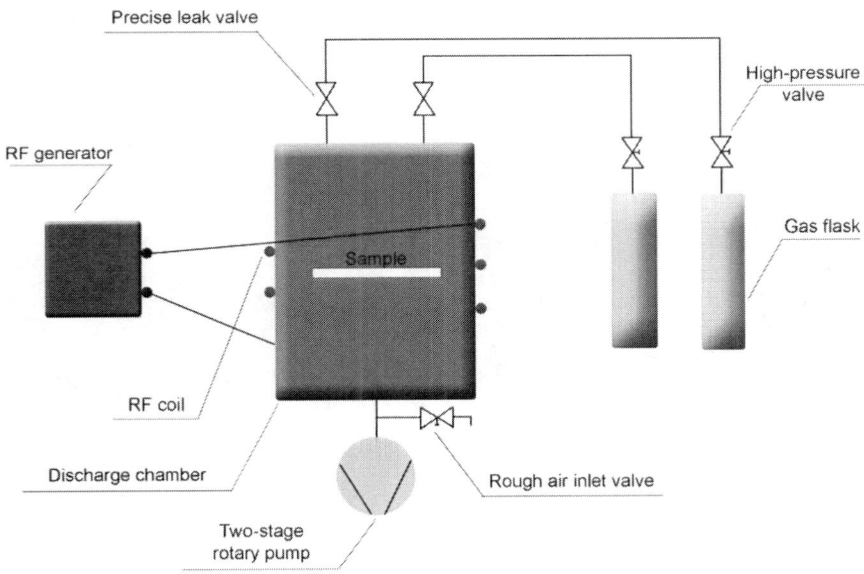

Figure 1. Schematic diagram of low-pressure RF plasma reactor

2.2. Morphological, Chemical and Physical Properties Of Untreated And Plasma-Treated Textiles

2.2.1. X-Ray Photoelectron (Xps) Analysis

Information on the chemical composition and chemical bonds of surface atoms of untreated and plasma-treated textile samples was obtained with XPS analysis. During the XPS analysis, a sample is illuminated with monochromatic X-ray light in an XPS spectrometer and the energy of emitted photoelectrons from the sample surface is analyzed. In the photoelectron spectrum, which represents the distribution of emitted photoelectrons as a function of their binding energy, peaks can be observed that may correspond to the elements present on the sample surface up to about 6 nm in depth. From the shape and binding energy of the peaks within XPS spectra, the chemical bonding of surface elements was inferred with the help of data from the literature.

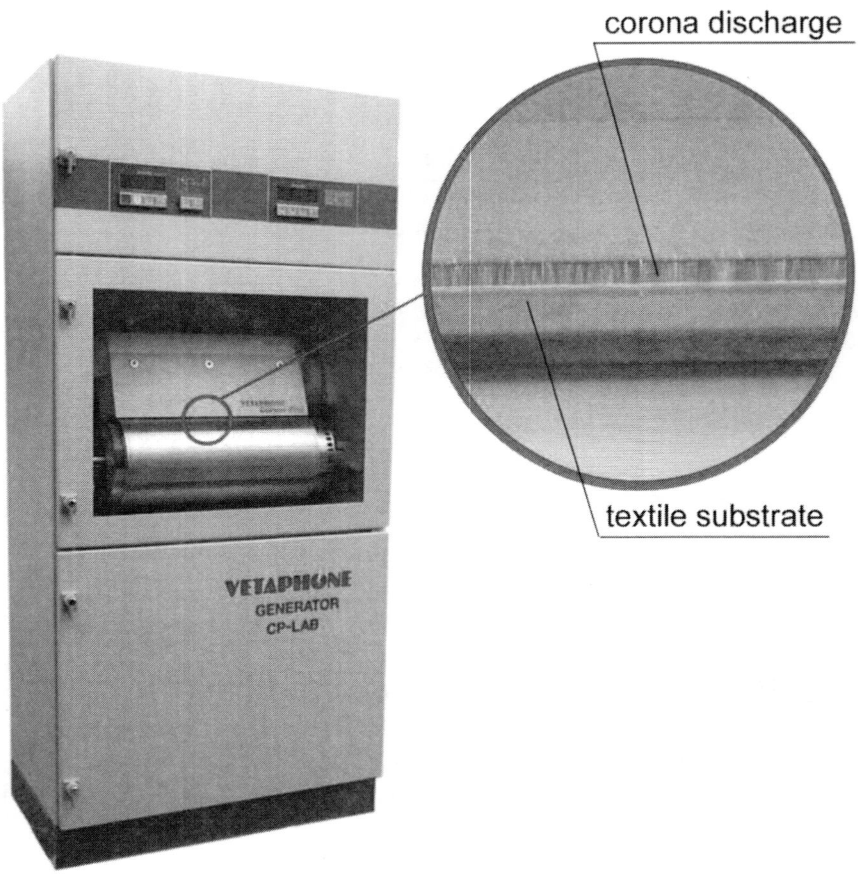

Figure 2. Picture of atmospheric corona plasma reactor and a detail of treating area shown in a red circle

2.2.2. Scanning Electron Microscopy (Sem)

The morphological surface properties of fibers and their changes after plasma treatment were studied using scanning electron microscopy (JEOL SEM type JSM-6060LV). All samples were coated with carbon and a 90% Au/10% Pd alloy layer.

2.2.3. Dynamometer Tensile Testing

Breaking strength and elongation of untreated and plasma-treated fabrics were analyzed according to the ISO 2062:1997 standard. Instron 6022 was used for this purpose. 100 mm samples of cotton yarn were analyzed using a pre-loading of 0.5 cN/tex and a speed of 250 mm/min. Samples were conditioned according to the ISO 139 standard. Tensile stress (cN/dtex) and elongation ε (%) are the mean values of the measured tensile strength and the elongation of 10 specimens, respectively, in the warp and in the weft directions.

2.3. Modification By Dyeing and Nanoparticles

2.3.1. Nanoparticles

For a research three different types of silver nanoparticles were used: silver nanoparticles of known dimensions (Ames Goldsmith Inc.), commercial form RucoBac AGP (Rudolph Chemie) and laboratory synthesized colloidal silver nanoparticles. Silver nanoparticles of known dimensions was medium density mono-dispersed 30 nm silver powder (Silver Nano Powder NP-30) and high density mono-dispersed 80 nm silver powder (Silver Nano Powder NP-80) [68]. The powdered silver nanoparticles are namely intended for a fine line printing (Ink-jet printing) and for electronics use. Because of their purity and known size they are very suitable and important for studies of adhesion to textile materials [69]. The initial concentration of silver powders in the dyeing baths was 20 mg/l. Commercial form of silver nanoparticles RucoBac AGP is a hygienic finish for all fiber types. It is a highly concentrated hygiene and freshness system complying with the Oeko-Tex® standard

100. RucoBac AGP is a nano-dispersion of titanium dioxide (TiO_2) as the carrier of the active component silver chloride (AgCl) [70]. The recommended concentration of RucoBac AGP is 0.2–0.5%. Laboratory synthesized colloidal silver nanoparticles were made by a reduction of silver salt in an aqueous solution at room temperature. An amount of 25.5 mg of $AgNO_3$ was dissolved in 750 ml of bi-distilled water, and argon gas was introduced into the solution for 30 min. The reduction was performed with 375 mg of $NaBH_4$ under constant stirring. The solution was kept under an argon atmosphere for another hour [71].

2.3.2. Dyeing of Cotton Samples

2.3.2.1. Dyeing with Reactive Dyes.

Exhaust dyeing with or without the addition of silver nanoparticles [3, 27, 72] was performed on untreated and plasma-treated cotton samples. The dye solutions were separately prepared using two different dyeing concentrations: blank (dyeing without dyestuff) and 0.05% on weight of fabric (owf) of Cibacron Deep Red S-B in liquor ratio 20 : 1. Dyeing baths contained 30 g/l of salt (Na_2SO_4anhydrous) and 8 g/l of sodium carbonate (Na_2CO_3 anhydrous). After the dyeing process, rinsing with distilled water was performed and neutralization with 1 ml/l of 30% acetic acid (CH_3COOH). As a soaping agent 1 g/l of CIBAPON R was used.

2.3.2.2. Vat Dyeing Procedure

Vat dyeing was performed on untreated and plasma-treated cotton samples. The dyed cotton fabrics were post-treated in a colloidal silver solution [71]. A dyeing bath was prepared with 4% owf Bezathren Blau BCE, 15 ml/l NaOH 38°Bé and 3.5 g/l $Na_2S_2O_4$. The liquor ratio was 83.3 : 1. The dyeing of cotton fabrics was performed at 60°C for 60 min. The dyed samples were rinsed twice in deionised water for 5 min, post-treated in 2 ml/l HCOOH 85% for 5 min and rinsed in deionised water for 5 min.

2.4. Color Measurements

Color measurements of differently modified samples were performed after conditioning them according to the ISO 139 standard. CIE standard illuminant D65/10 was used on a Datacolor Spectraflash SF 600-CT reflection spectrophotometer. The aperture diameter of the measuring port of the spectrophotometer was 9 mm.

2.5. Elemental and Antimicrobial Analysis

The quantity of adsorbed silver was determined using ICP-MS. This technique combines a high-temperature inductively coupled plasma (ICP) source with a mass spectrometer (MS). The antibacterial efficiency of functionalized textiles was determined using microbiological tests according to the ASTM Designation: E 2149–01 method. Antibacterial activity of the cotton fabrics was tested by a certified laboratory against *Staphylococcus aureus* (ATCC 25923), *Escherichia coli* (ATCC 25922), *Streptococcus faecalis* (ATCC 27853) and *Pseudomonas aeruginosa* (ATCC 27853).

3. RESULTS AND DISCUSSION

The influence of different plasma systems on the adhesion of nanoparticles is discussed. The emphasis of the study is to use minimal concentrations, initially, of nanoparticles for loading onto textiles and to achieve maximum quantity on the material. Exhaust dyeing process was used for loading of silver nanoparticles onto textiles. Before applying nanoparticles to any material, its surface needs to be adequately prepared and chemically and morphologically well analyzed. Only good conditions on the substrate surface can provide a qualitative deposition of particles [73]. In literature one can find quotations of XPS analysis of plasma modified cotton substrates which were pre-prepared with various procedures prior to plasma modification (i.e. alkaline boiling, scouring, laundering). But to study plasma modification of cotton it is important to know about the surface changes of substrates that were not cleaned or otherwise pre-prepared. Therefore, the chemical surface changes were evaluated for raw, bleached and bleached/mercerized cotton before and after plasma treatment.

The surface of raw untreated cotton fabric contains a high concentration of carbon and a low concentration of oxygen. This is not characteristic for native cellulose [74]. XPS spectrum C 1s of cellulose also does not include C-C/C-H bonds (Figure 3). The surface of raw cotton is rich in C atoms (C-C/C-H bonds), what could indicate the presence of C-C/ C-H bond rich substances, such as waxes, pectin and proteins. The surfaces of bleached and bleached/mercerized cotton fabrics are alike but very different from raw cotton. Bleached and bleached/mercerized cotton samples are pre-oxidized due to the scouring, bleaching and mercerizing process, and therefore more similar to native cellulose. After modifications with atmospheric air corona plasma and water vapor low-pressure plasma cotton samples contain a higher O/C ratio (the samples contain more of the oxygen and less of the carbon atoms) (Figure 4).

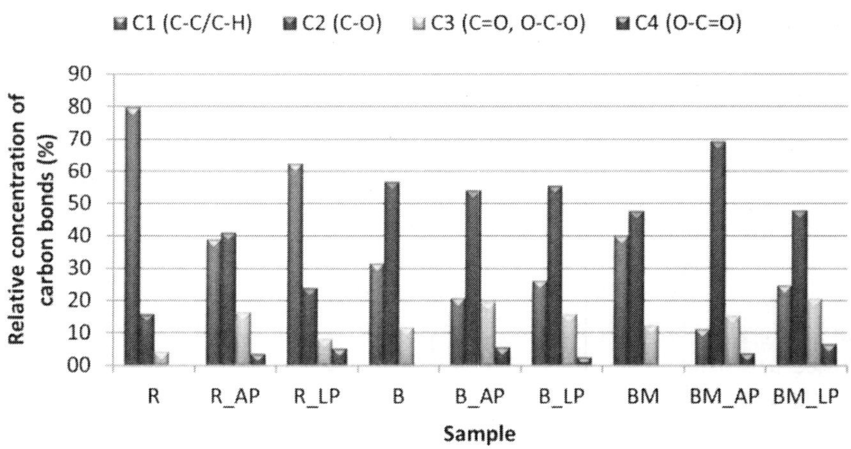

Figure 3. Relative concentration of carbon bonds on the surface of cotton samples: R – raw cotton, B – bleached cotton, BM – bleached/mercerized cotton, AP – atmospheric plasma treatment, LP – low-pressure plasma treatment

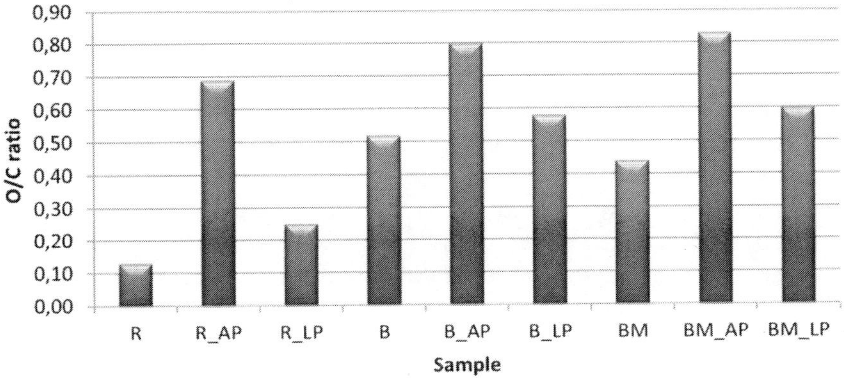

Figure 4. The concentration ratio between oxygen and carbon on the surface of cotton samples: : R – raw cotton, B – bleached cotton, BM – bleached/mercerized cotton, AP – atmospheric plasma treatment, LP – low-pressure plasma treatment

The increase of O/C ratio is expected since plasma interaction with cotton causes the surface oxidation. Changes of C-atom bonds on raw cotton samples after corona and low-pressure plasma treatment are visible in Figure 3 as the ratio of C-O and C=O bonds increased. The results show that increase of relative concentration of C-O bonds is distinctive after corona plasma treatment (41 %) than after low-pressure plasma treatment (23.9 %). It is similar for C=O bonds. The relative concentration of C=O bonds on the surface of raw cotton increases to 16.6 % after corona plasma treatment and to 8.2 % after low-pressure plasma treatment. Both plasma treatments result in a formation of O-C=O bonds. C 1s spectrum of plasma-treated samples resembles to the spectra of native cellulose, which could indicate that plasma selectively removed the noncellulosic parts present on a surface of a raw untreated cotton fabric. After plasma treatment the oxygen content on surface of bleached and bleached/mercerized cotton samples increases. After a treatment with corona plasma the content of oxygen on bleached cotton sample increases from 34.3 at% to 44.5 at%, while the treatment with low-pressure plasma increases oxygen to 36.8 at%. After both plasma treatments the content of C-O remains the same (~55 %), the content of C=O bonds increases and appearance of O-C=O bonds is noticeable. Increase of bonds is distinctive after corona plasma treatment than after low-pressure plasma treatment. Content of C-C/C-H bonds decreases after both plasma treatments. Increase of O/C ratio on plasma-treated bleached/mercerized cotton samples is distinctive after corona plasma treatment, where the oxygen concentration increases to 45.3 at%, while increases to 37.5 at% after low-pressure plasma treatment. Due to oxidation process in plasma, the content of C-C/C-H bonds deceases and the change is more noticeable for bleached/mercerized cotton than for a bleached cotton fabric. For bleached/mercerized cotton sample concentration of C-O bonds is increased in the case of corona plasma treatment, from 47.6 % to 69.3 %, while it remains almost equal after low-pressure plasma treatment (47.8 %). After both plasma treatments the content of C=O bonds increases and appearance of O-C=O bonds is noticeable. These changes are more distinct after low-pressure plasma treatment (20.7 % and 6.8 %) than after corona plasma treatment (15.5 % and 3.9 %). The results summarized in Figure 4 show higher concentration of oxygen on atmospheric air corona and water vapor low-pressure plasma-treated raw cotton than on untreated bleached or bleached/mercerized cotton. From these results it can be concluded that by using plasma technology some of the technological processes of pretreatment of cotton fabrics before dyeing can be avoided. Plasma-treated raw cotton fabric absorbs the same or more of dyestuff as untreated bleached or bleached/mercerized cotton fabric [3, 65, 65, 76]. Reactive dyes are typical anionic dyes containing reactive systems which in alkaline media make bonds with –OH groups of cotton fibers. Increase of dyeability of cotton treated with atmospheric air corona plasma or water vapor low-pressure plasma is correlated to an increase of the number of hydrophilic groups on the surface of the fibers. In Figure 5 the correlation between color differences (expressed as ΔE^*) of reactive dyed cotton and differences in O/C ratio (expressed as $\Delta O/C$) of untreated and plasma-treated cotton is presented. The most perceptible changes in color differences (ΔE^*) are noticeable when differences in O/C ratio ($\Delta O/C$) are higher. Visible color differences between untreated and water vapor plasma-treated cotton were noticeable for raw ($\Delta E^* = 2.20$) and bleached/mercerized cotton ($\Delta E^* = 1.07$).

The samples modified in tetrafluoromethane low-pressure plasma (CF_4 plasma) gave different results. There was no visible color difference between untreated and CF_4 plasma-treated bleached/mercerized cotton samples ($\Delta E^* = 0.62$). Both samples were practically equally colored. Treating cotton fabrics for 10 sec with

CF_4 plasma does not influence their dyeing properties. The results of XPS study show a small increase in the concentration of oxygen after CF_4 plasma treatment, but the result is hardly remarkable (Figure 6 and 7).

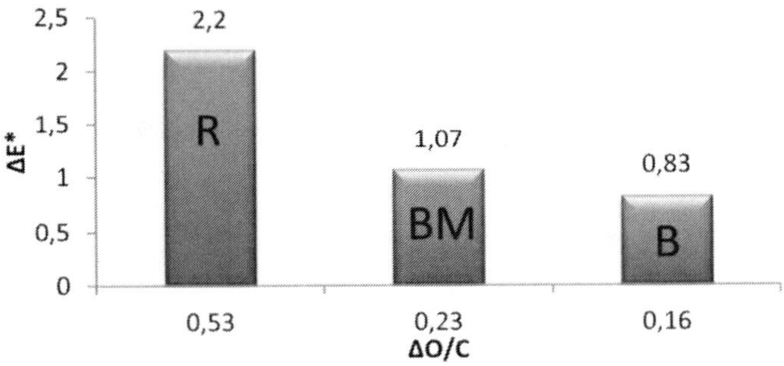

Figure 5. The correlation between color differences (expressed as ΔE*) of reactive dyed cotton and differences in O/C ratio (expressed as ΔO/C) of untreated and plasma-treated cotton (R – raw cotton, B – bleached cotton, BM – bleached/mercerized cotton)

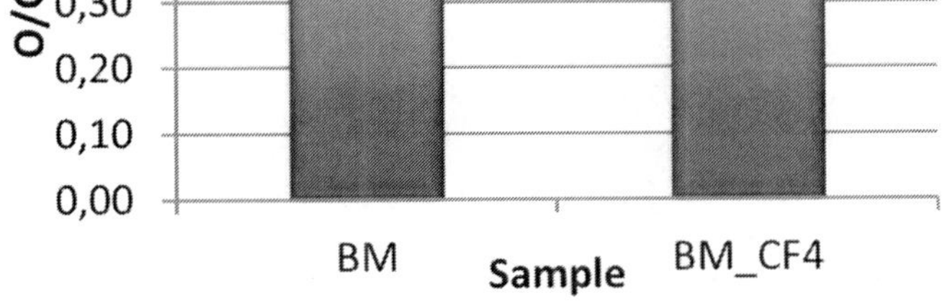

Figure 6. The concentration ratio between oxygen and carbon on the surface of untreated (BM) and CF_4plasma-treated (BM_CF4) cotton samples

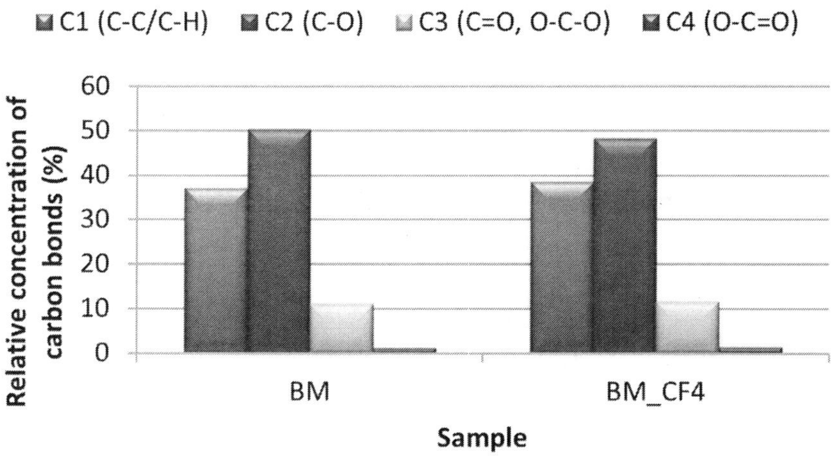

Figure 7. Relative concentration of carbon bonds on the surface of untreated (BM) and CF_4 plasma-treated (BM_CF4) cotton samples

It is interesting, however, that no fluorine was observed in the XPS spectra, but that the additional ISE analysis indicated the presence of <5 ppm of total fluoride on CF_4 plasma-treated samples. Here, it is worth mentioning that we are not the first group to observe little or no fluorine on organic materials treated with CF_4 plasma [76, 77]. Although XPS analysis did not show specific chemical surface differences between untreated and CF_4 plasma-treated cotton samples, the SEM analysis showed strongly modified surface morphology of plasma-treated cotton (Figure 8), while smaller changes of the surface morphology of cotton samples treated with atmospheric air corona plasma and water vapor low-pressure plasma are noticed.

SEM image of untreated cotton (Figure 8 a) shows a typical grooved surface morphology with macrofibrils oriented predominantly in the direction of the fiber axis. The outlines of the macrofibrils are still visible, and they are smooth and distinct due to the presence of an amorphous layer covering the fiber. The surface of corona plasma-treated cotton has striped, cleaned and more distinct macrofibrilar structure (Figure 8 b). The same effect can be noticeable on the surface of cotton fibers treated with water vapor low-pressure plasma (Figure 8 c). The plasma-treated fiber surface remains grooved and the macrofibrile structure has gained a much sharper outline. The individual macrofibrils (0.2–1 nm) and their transversal connections are visible in the primary cell wall. Between them, narrow voids thinner than 10 nm are noticeable. The surface morphology of CF_4 plasma-treated cotton (Figure 8 d) is very different comparing to those treated with water vapor low-pressure or corona plasma. CF_4 plasma-treated cotton has an extremely rough and nanostructured surface with dimensions of the grains roughly between 150 and 500 nm. Dissociation energies of plasma molecules are the reason for such rich etched surface. Both CF_4 and H_2O molecules get dissociated in our plasma [78]. The dissociation energy of CF_4 is 12.6 eV [79] while the ionization energy of CF_4 is about 16 eV, which is much higher than the dissociation energy of water vapor or OH molecules, which is at about 5 and 4 eV, respectively. The electrons with energy of few eV are therefore likely to dissociate water and OH molecules rather than dissociate the CF_4 molecules. The result is an extremely high dissociation fraction of water molecules and a moderate dissociation fraction of CF_4 molecules. Since the partial pressures of water and CF_4 are comparable, it can be concluded that the density of O and OH radicals in our plasma is at least as high as the density of CF_x radicals if not more so. The textile sample exposed to plasma is therefore subjected to interact with CF_x, O, OH, F and H radicals. The density of H is probably as high as that of F so extensive recombination to HF is expected. CF_xradicals tend to graft onto the textile, but the grafting probability on cellulose is low compared to the interaction probability with O and OH radicals. O and OH radicals are extremely reactive and cause etching of cellulose. Fluorine atoms are efficient at abstracting hydrogen in the first step of the oxidation reaction. In addition, the presence of fluorine atoms in the plasma enhances the dissociation of the oxygen, further increasing the ashing rates [80]. CF_4 is the gas most commonly added to oxygen to enhance the generation of atomic oxygen in plasma and to increase polymer etch rates [81].

While the surface of plasma-treated cotton was changed, the mechanical properties of cotton fabrics did not alter after plasma modification. The breaking strength and elongation of textiles practically do not change (Figure 9 and 10) [3, 27]. These results are in accordance with the results obtained by other authors [82-85]. Plasma modification of textiles does not impair their mechanical properties.

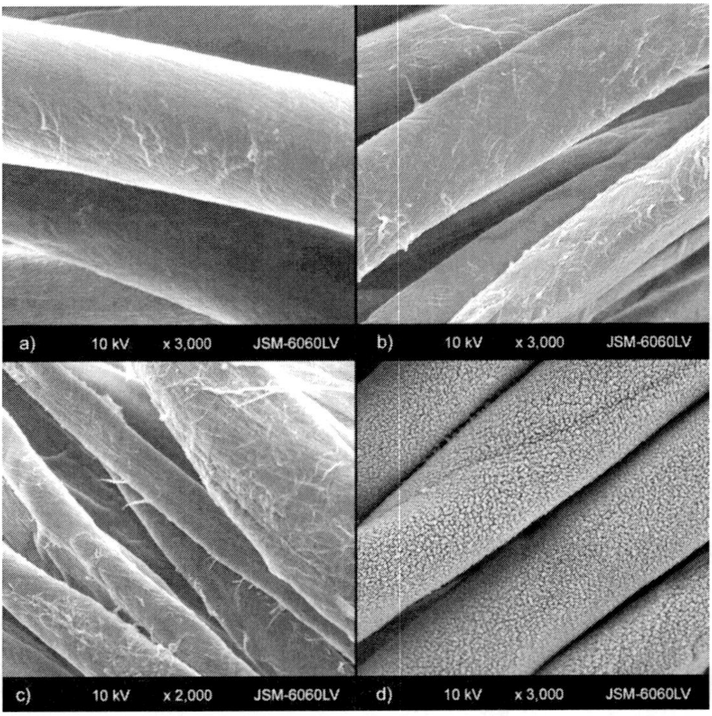

Figure 8. SEM images of untreated and plasma-treated cotton samples

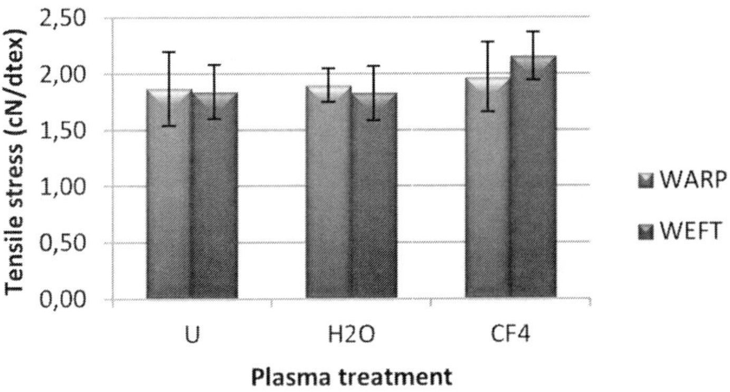

Figure 9. Tensile stress (cN/dtex) of cotton after plasma treatment: U – untreated, H2O - ICRF water vapor plasma-treated, CF4 - ICRF tetrafluoromethane plasma-treated

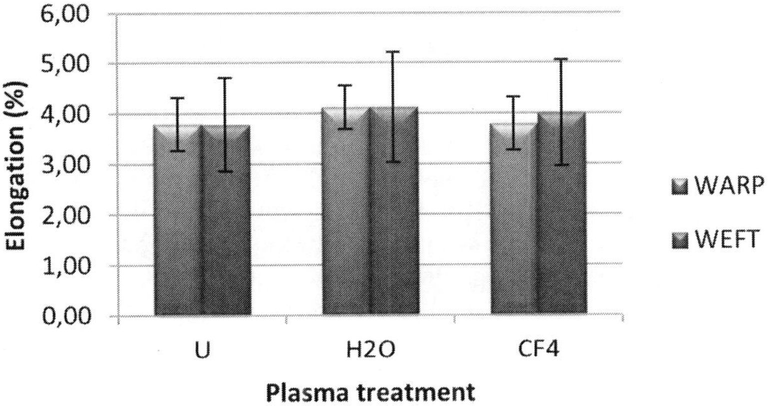

Figure 10. Elongation (%) of cotton after plasma treatment: U – untreated, H2O - ICRF water vapor plasma-treated, CF4 - ICRF tetrafluoromethane plasma-treated

A textile surface with such extremely rich surface morphology (as seen from SEM) is likely to influence higher adsorption of silver nanoparticles, which is in accordance with our ICP-MS results inTable 1, where the quantity of adsorbed silver onto cotton samples is presented.

From the results summarized in Table 1, it is clear that 30 nm silver nanoparticles are more adhesive to the cotton fabric than 80 nm silver nanoparticles. Regarding their volume, nanoparticles have a high specific surface area. By decreasing the size of a particle, its surface distribution rate increases [87]. The quantity of adsorbed silver on untreated bleached/mercerized cotton was 32 ppm when 30 nm silver nanoparticles were used and 13 ppm when 80 nm silver nanoparticles were used. Similar trend can be observed when cotton was modified with plasma. When cotton was treated with water vapor low-pressure plasma the quantity of adsorbed 30 nm silver nanoparticles was 50 ppm and the quantity of adsorbed 80 nm silver nanoparticles was 17 ppm, which means that adsorption of 80 nm silver nanoparticles was three times lower than the adsorption of 30 nm silver nanoparticles. The results show that plasma heavily modifies the morphology and surface chemical properties of cotton and by that has a great impact on the adsorption of silver nanoparticles onto fabrics. Atmospheric air corona treatment of cotton fabrics enhanced the quantity of silver onto raw cotton up to 4 times and onto bleached/mercerized cotton up to 2 times [64]. The adsorption of 80 nm silver nanoparticles onto CF_4plasma-treated cotton was 2 times higher and onto water vapor plasma-treated cotton was 1.3 times higher than on untreated cotton. These fabrics had a sufficient antimicrobial effectiveness against*Escherichia coli* and *Pseudomonas aeruginosa* (Table 1) [3]. The adsorption of 30 nm silver nanoparticles onto CF_4 plasma-treated cotton is 1.7 times higher and onto water vapor plasma-treated cotton was 1.6 times higher than on untreated cotton. This gave a good antimicrobial effectiveness against *Enterococus faecalis* and *Pseudomonas aeruginosa* (Table 1) [27].

Table 1. The quantity of silver (ppm) and antimicrobial efficiency expressed as a bacterial reduction (%) of untreated and plasma-treated cotton samples functionalized with powdered silver nanoparticles

Plasma modification of cotton	Functionalization with silver nanoparticles	Ag quantity (ppm) on cotton	Bacterial reduction (%)
Untreated	30 nm	32	–[a]
	80 nm	13	–[a]
Corona air	80 nm	39	–[a]
ICRF water vapor	30 nm	50	52 (*E. coli*) 64 (*P. aeruginosa*)
	80 nm	17	–[a]
ICRF CF_4	30 nm	54	68 (*E. faecalis*) 77 (*P. aeruginosa*)
	80 nm	26	

–[a] no bacterial reduction

Although the purpose of our research was to study the efficiency and appropriateness of plasma modification of textiles in order to achieve a higher adsorption of nanoparticles onto their surfaces, the wash fastness of functionalized textiles was examined as well. Wash fastness test was carried out in a laboratory apparatus Launder-O-Meter [3, 27]. The samples were washed repetitively ten times at 95°C in a solution of 5 g/l of SDC standard detergent and 2 g/l of Na_2CO_3 (where 10 globules were added). The duration of the washing cycles was 30 min. After every wash cycle, the samples were rinsed twice in distilled water and then for 10 minutes under a tap water, which was followed by squeezing and air drying. In Table 2 the results of silver quantity on washed cotton samples are presented. From the results it can be seen that after ten times washing at 95°C the quantity of silver on the cotton samples decreased (Table 2).

Table 2. The quantity of silver (ppm) before and after washing

Plasma modification of cotton	Functionalization with silver nanoparticles	Ag quantity (ppm) on cotton before washing	Ag quantity (ppm) on cotton after washing
ICRF water vapor	30 nm	50	26
	80 nm	17	10
ICRF CF_4	30 nm	54	36
	80 nm	26	22

When cotton was modified by water vapor plasma the quantity of 30 nm silver nanoparticles decreased by 48%, and when modified by CF_4 plasma the quantity of silver nanoparticles decreased by 34%. The drop in the concentration of silver on the cotton after ten wash cycles is also observed with the 80 nm silver nanoparticles. Cotton that is treated with water vapor plasma loses 41%; however, cotton treated with CF_4 plasma loses 15% of the silver nanoparticles. After a 10-second treatment with CF_4 plasma, the surface of the bleached/mercerized cotton is more liable to the adsorption of silver nanoparticles than the surface that was modified by the water vapor plasma. The particle adhesion is a complex phenomenon depending on the interaction mechanism between particles and the surface of a material. Adsorption of silver nanoparticles onto cotton from a dyeing bath is much higher for smaller particles than for larger ones. When particles are already on the surface of cotton fibers the attractive interactions and forces are stronger for 80 nm silver nanoparticles than for 30 nm silver nanoparticles. That is the reason for a better wash durability of 80 nm silver particles. From the results presented inTable 1 and 2, it is evident that the quantity of silver on untreated and unwashed cotton is almost the same as on the samples that were modified by plasma and washed 10 times at 95°C.

The further experiments in functionalization of cotton with other forms of silver nanoparticles (i.e. commercial form RucoBac AGP and laboratory synthesized colloidal silver nanoparticles) were based on the obtained adhesion results of silver nanoparticles of known dimensions. The application of RucoBac AGP and synthesized colloidal silver nanoparticles was carried on a bleached/mercerized cotton fabric modified by atmospheric air corona plasma and water vapor low-pressure plasma.

The exhaustion method was used for the deposition of RucoBac AGP onto blank dyed, dyed, plasma-treated blank dyed and plasma-treated dyed cotton samples. 0.1 % of RucoBac AGP was used, which represents a half of the lowest concentration recommended by the agent producer. The reason for such a decision was to verify the possibility in achieving good antibacterial efficiency of a cotton fabric with the use of a very low concentration of silver composite. The liquor ratio was 10 : 1, and the treatment time 30 min at 50 ºC. Afterwards, the samples were dried at 130 ºC for 4 min. The results of the ICP-MS and antimicrobial analysis of samples are presented in Table 3.

Table 3. The quantity of silver (ppm) and antimicrobial efficiency, expressed as a bacterial reduction (%), of untreated and plasma-treated cotton samples functionalized with RucoBac AGP

Sample Treatment	Plasma treatment of cotton	Ag quantity (ppm) on cotton	Bacterial reduction (%)
Functionalization with 0.1 % RucoBac AGP	Untreated	1,5	—[a]
	Corona air	1,8	—[a]
	ICRF water vapor	11	58 (*E.coli*) 98 (*S.faecalis*)
Dyeing with reactive dye and functionalization with 0.1 % RucoBac AGP	Untreated	8,9	100 (*E.coli, S.aureus, S.faecalis* and *P.aeruginosa*)
	Corona air	8,4	100 (*E.coli, S.aureus, S.faecalis* and *P.aeruginosa*)
	ICRF water vapor	13	80 (*S.aureus* and *E.coli*) 100 (*S.faecalis*)

—[a] no bacterial reduction

The results summarized in Table 3 show low adsorption of silver onto untreated blank dyed cotton. The adsorption of silver onto corona modified cotton did not significantly increase. Both samples do not exhibit the antimicrobial effectiveness. However, treating cotton with water vapor low-pressure plasma increased the adsorption of silver onto cotton up to 7 times, which resulted in an excellent antimicrobial effectiveness against *Streptococcus faecalis* and a sufficient antimicrobial effectiveness against *Escherichia coli*. The adsorption of silver significantly increased when RucoBac AGP was applied onto dyed cotton fabric, regardless of plasma treatment. RucoBac AGP is a nano-dispersion of TiO_2 as the carrier of the active component AgCl. In the presence of moisture, silver cations react with hydroxyl functional cellulosic groups and are attached to each other electrostatically. The presence of a reactive dye on cotton and the introduction of additional covalently bound sulfonic acid groups will facilitate the uptake of a cationic antimicrobial agent [87]. Therefore, it is possible to conclude that RucoBac AGP is bound to the cotton surface through sulfonic groups of a covalently bound dye and through partially ionized hydroxyl and carboxyl groups present on the fiber [72]. The significant increased adsorption of silver can be also noticed with samples modified by water vapor low-pressure plasma and dyeing, while there was no noticeable change when samples were modified by corona plasma. Results obtained from XPS analysis showed that plasma modification of bleached/mercerized cotton increased the content of oxygen on the surface. In addition, after both plasma treatments the appearance of new bonds was noticeable (O-C=O) and the content of C=O bonds increased. That means that oxygen rich functional groups incurred on the surface of plasma modified cotton. These changes were more distinct after low-pressure plasma treatment than after corona plasma treatment. The difference in both plasma treatments is noticeable from the ICP-MS results also (Table 3), since the adsorption of RucoBac AGP onto cotton was greater after modification by water vapor low-pressure plasma. Nevertheless, all modified cotton fabrics had an excellent antimicrobial effectiveness against *Staphylococcus aureus*, *Escherichia coli*, *Streptococcus faecalis* and *Pseudomonas aeruginosa*.

The objectives for textile industry have been continuous on-line treatments of fabrics. Low-pressure plasma reactors, such as radio frequency powered plasma, provide greater stability and uniformity but generally require more handling of textile materials through the vacuum system than corona discharges at atmospheric pressures. In this respect, the use of corona plasma is more appropriate for the industry. In our following research when laboratory colloidal silver was used, we focused on treating cotton fabric with atmospheric corona plasma. The synthesis of colloidal silver was performed by reducing silver salt in an aqueous solution at room temperature under argon atmosphere. The procedure was described previously, in section 3.2.1. A synthesis of colloidal silver and loading of silver onto cotton fabric was performed as the second phase after dyeing cotton fabrics with a blue vat dye. To verify whether vat-dyeing influences the adsorption of colloidal silver onto cotton fabric, a blank vat-dyeing procedure (dyeing with all chemicals and no dye) was also performed [71]. Apart from indigo, the vat dyes used in dyeing applications are mainly derivatives of anthraquinone and of higher condensed aromatic ring systems with a closed system of conjugated double bonds. They generally contain two, four or six reducible carbonyl groups [88].

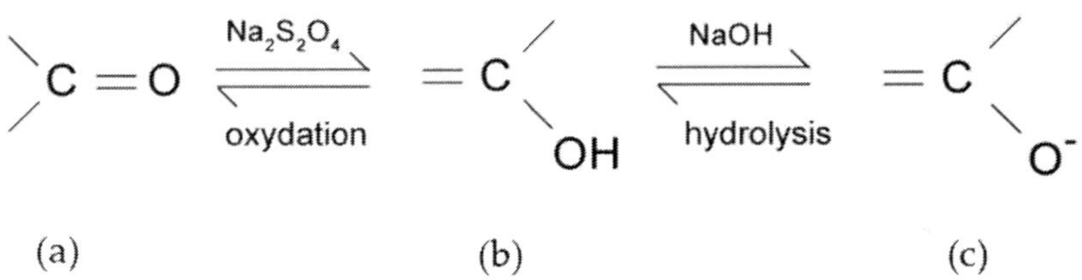

Figure 11. Scheme of reduction and oxidation process of carbonyl group of vat dye

The vat dyes are insoluble in the keto form (Figure 11a). For dyeing they must be transformed to a water soluble enolate (leuko) form (Figure 11c) by a reduction. This form of the dye is appropriate for cellulose dyeing, but the addition of electrolyte is also required. After the dyeing, the dye in amorphous regions is transformed through leuko acid (Figure 11b) into its original water insoluble form by rinsing and oxidation [89]. The vat dyeing of bleached/mercerized cotton fabric before colloidal silver treatment significantly influences the adsorption of silver onto cotton fabrics [71]. Also the adsorption of silver is influenced by an immersion time of cotton into colloidal silver solution. Although it was proven that modification of cotton by corona plasma strongly influences the adsorption of powdered silver nanoparticles by increasing their quantity on the modified cotton, the experiment using synthetized colloidal silver on corona treated cotton had to be conducted. The Table 4 presents the results of ICP-MS and antimicrobial analysis of untreated, corona and vat dyed colloidal silver loaded cotton.

Table 4. The quantity of silver (ppm) and antimicrobial efficiency, expressed as a bacterial reduction (%), of untreated and plasma-treated cotton samples functionalized with synthetized colloidal silver

Sample description	Ag quantity (ppm) on cotton	Bacterial reduction (%)
Dyed and functionalized with colloidal silver	24	100 (*E.coli*, *S.aureus*, *S.faecalis* and *P.aeruginosa*)
Corona plasma-treated, dyed and functionalized with colloidal silver	43	100 (*E.coli*, *S.aureus*, *S.faecalis* and *P.aeruginosa*)
Corona plasma-treated and functionalized with a half of concentration of colloidal silver	1.6	—[a]
Corona plasma-treated, dyed and functionalized with a half of concentration of colloidal silver	4.6	100 (*E.coli*, *S.aureus*, *S.faecalis* and *P.aeruginosa*)

—[a] no bacterial reduction

The results summarized in Table 4 show that atmospheric air corona treatment enhanced the quantity of silver onto dyed cotton up to 1.8 times comparing to the untreated dyed cotton. In addition to the morphological changes induced by plasma (seen by SEM), XPS analysis showed the increase of C-O and C=O bonds and formation of O-C=O bonds on the surface of treated cotton. The increased concentration of oxygen and newly formed bonds contributed to a better adhesion of Ag^+ ions from the colloidal solution onto cellulosic fibers. In addition to the increased number of functional groups containing oxygen, the dyed cotton fabric contains additional anionic sites due to the partial ionization of the molecules of insoluble vat dye. Colloidal silver is produced in a water solution using $AgNO_3$reduced by $NaBH_4$. $NaBH_4$ also slightly reduces the water insoluble vat dye on the dyed fabric into a slightly soluble form. Although $NaBH_4$ is not a sufficiently strong reducing agent for dyeing cotton with a vat dye [90], it is nevertheless strong enough to enhance the negative charge of the dye on the fabric [91] such that silver ions (Ag^+) from a silver colloidal solution can be electrostatically attracted to the dyed cotton surface. It was reported that while Ag^+ ions from a colloidal silver solution are exhausted to the anionic cotton fibers to a high degree because of the attractive electrostatic interactions, the high increase of the adsorption ability of silver nanoparticles caused by the van der Walls forces resulted from the high surface area to volume ratio of these particles [43].

The important goal of our research was to use minimal concentrations, initially, of silver nanoparticles for loading onto textiles and to achieve a maximum quantity on the material, and thus to achieve functionalized cotton textile with an excellent antimicrobial efficiency. Since the cotton fabrics already had an excellent antimicrobial efficiency when functionalized with rather low concentration of silver, the decision was made to verify the possibility in achieving good antibacterial efficiency of a cotton fabric with the use of half of the initial concentration of $AgNO_3$ and $NaBH_4$. In this case, the corona plasma modified and undyed cotton contained 1.6 ppm of silver. The quantity of silver was so low that the fabric did not have an antimicrobial efficiency. However, when cotton fabric was modified by corona plasma and then vat dyed the quantity of silver was 4.6 ppm, which gave the fabric an excellent antimicrobial efficiency against *Staphylococcus aureus*, *Escherichia coli*, *Streptococcus faecalis* and*Pseudomonas aeruginosa*. The antimicrobial analysis showed the antimicrobial ineffectiveness of dyed cotton sample, therefore the dye itself did not contribute to the antibacterial efficiency of the functionalized cotton sample.

4. CONCLUSION

Our research shows that plasma treatment is an effective method to be used in achieving surface changes on the textile material by changing the functional groups on the textile surface and by changing the morphology of the fibers. The results of adsorption of different forms of silver nanoparticles on untreated and plasma-treated surfaces of fabrics confirm the fact that, for nanotechnological processes, the surface of the material has to be properly prepared. The adsorption of metal nanoparticles on textile materials depends on specific chemical and morphological properties of fibers. Plasma modification of cotton had a positive influence on the increased adsorption of silver nanoparticles loaded during exhaust dyeing process. From the bath, which contained a low concentration of silver nanoparticles, we have successfully applied a greater quantity of silver onto plasma modified cotton (up to four times). In some cases using plasma the dyeability of cotton was also improved. We succeeded to create a cotton fabric containing minimal quantity of silver with an excellent antimicrobial efficiency. This is very interesting from the technological point of view since by this method the quantity of silver in wastewater can be dramatically reduced. Another important result of the research was that plasma modification did not impair the mechanical properties of textiles. By using plasma technology new and improved properties of materials can be created that cannot be achieved by standard procedures, where nanostructuring of natural and synthetic fibers is emphasized. The use of plasma for modification of textiles brings to the textile industry many novelties, since plasma technology can be used as a substitute or as a support to the existing technologies, and by that positively influences the economy and ecology of the industrial processes. The knowledge of using plasma technology enables an introduction of contemporary (state-of-the-art) and ecological process of a textile modification into the textile industry and a development of highly technological products with improved or new properties.

ACKNOWLEDGEMENTS

This work was financially supported by Slovenian Research Agency (Programme P2-0213 Textile and Ecology), and Eureka Nanovision project.

REFERENCES

1. Kang J.Y., Sarmadi M. Textile Plasma Treatment Review – Natural Polymer-Based Textiles. AATCC Review 2004; 4(10) 28-32.

2. Gorjanc M., Recelj P., Gorenšek M. Plasma Technology for Textile Purposes. Tekstilec 2007; 50(10-12) 262-266.

3. Gorjanc M., Bukosek V., Gorensek M., Vesel A. The Influence of Water Vapor Plasma Treatment on Specific Properties of Bleached and Mercerized Cotton Fabric. Textile Research Journal 2010; 80(6) 557-567.

4. Mozetic M., Cvelbar U., Sunkara M.K., Vaddiraju S. A Method for the Rapid Synthesis of Large Quantities of Metal Oxide Nanowires at Low Temperatures. Advanced Materials 2005; 17(17) 2138-2142.

5. Levchenko I., Rider A. E., Ostrikov K. Control of Core-shell Structure and Elemental Composition of Binary Quantum Dots. Applied Physics Letters 2007; 90(19) 193110.

6. Drenik A., Vesel A., Mozetič M. Controlled Carbon Deposit Removal by Oxygen Radicals. Journal of Nuclear Materials 2009; 386-88, 893-895.

7. Huang S.Y., Arulsamy A.D., Xu M., Xu S., Cvelbar U., Mozetič M. Customizing Electron Confinement in Plasma-Assembled Si/AlN Nanodots for Solar Cell Applications. Physics of Plasmas 2009; 16(12) 123504-1-123504-5.

8. Mozetic M. Reactive Plasma Technologies in Electronic Industry, Informacije MIDEM 2003; 33(4) 222-227.

9. Ferreira J.A., Tabares F.L., Tafalla D. Removal of Carbon Deposits in Narrow Gaps by Oxygen Plasmas at Low Pressure. Journal of Vacuum Science and Technology A 2007; 25(4) 746-750.

10. Shishoo R. Plasma Conquering the Textile Industry. Magazine for European Research 1999; 24, 22-23.

11. Fridman, A. Plasma chemistry. New York : Cambridge University Press. 2008.

12. Morent R., De Geyter N., Verschuren J., De Clerck K., Kiekens P., Leys C. Non-thermal Plasma Treatment of Textiles. Surface and Coatings Technology 2008; 202, 3427-3449.

13. Franceschini D.F., Achete C.A., Freire F.L., Beyer W., Mariotto G. Structural Modifications in a-c-h Films Doped and Implanted With Nitrogen. Diamond and Related Materials 1994; 3(1-2) 88-93.

14. Chen P., Wang J., Wang B., Li W., Zhang C., Li H., Sun B. Improvement of Interfacial Adhesion for Plasma-Treated Aramid Fiber-Reinforced Poly(Phthalazinone Ether Sulfone Ketone) Composite and Fiber Surface Aging Effects. Surface and Interface Analysis 2009; 41(1) 38-43.

15. Guo F., Zhang Z.-Z., Liu W.-M., Su F.-H., Zhang H.-J. Effect of Plasma Treatment of Kevlar Fabric on the Tribological Behavior of Kevlar Fabric/Phenolic Composites. Tribology International 2009; 42(2) 243-249.

16. Jia C., Chen P., Li B., Wang Q., Lu C., Yu Q. Effects of Twaron Fiber Surface Treatment by Air Dielectric Barrier Discharge Plasma on the Interfacial Adhesion in Fiber Reinforced Composites. Surface and Coatings Technology 2010; 204(21-22) 3668-3675.

17. Liu L., Jiang Q., Zhu T., Guo X., Sun Y., Guan Y., Qiu Y. Influence of Moisture Regain of Aramid Fibers on Effects of Atmospheric Pressure Plasma Treatment on Improving Adhesion with Epoxy. Journal of Applied Polymer Science 2006; 102(1) 242-247.

18. Bhoj A.N., Kushner M.J. Repetitively Pulsed Atmospheric Pressure Discharge Treatment of Rough Polymer Surfaces: I. Humid Air Discharges. Plasma Sources Science and Technology 2008; 17(3), art. no. 035024.

19. KinlocH A.J. The Science of Adhesion: Part 1: Surface and Interfacial Aspects. Journal of Materials Science 1980; 15, 2141-2166.

20. Wu S. Polymer Interface and Adhesion. New York : Marcel Dekker. 1982.

21. Marsh D.H., Riley D.J., York D., Graydon A. Sorption of Inorganic Nanoparticles in Woven Cellulose Fabrics. Particuology 2009; 7(2) 121-128.

22. Bozzi A., Yuranova T., Guasaquillo I., Laub D., Kiwi J. X Self-cleaning of Modified Cotton Textiles by TiO2 at Low Temperatures under Daylight Irradiation. Journal of Photochemistry and Photobiology A: Chemistry 2009; 174, 156-164.

23. Cunko R., Varga K. Application of Ceramics for the Production of High-Performance Textiles. Tekstil 2006; 55(6) 267-278.

24. Gorensek M., Gorjanc M., Bukosek V., Kovac J., Jovancic P., Mihailovic D. Functionalization of PET Fabrics by Corona and Nano Silver. Textile Research Journal 2010; 80(3) 253-262.

25. Gorenšek M., Gorjanc M., Bukošek V., Kovač J., Petrović Z., Puač N. Functionalization of Polyester Fabric by Ar/N2 Plasma and Silver. Textile Research Journal 2010; 80(16) 1633-1642.

26. Gorjanc M., Mozetič M., Gorenšek M. Low-temperature plasma for pretreatment of cotton fabric for better adhesion on nano silver. In: Simončič, B., Gorjanc, M. (eds.). Conference proceedings 19. June 2009. Ljubljana; 2009.

27. Gorjanc M., Bukošek V., Gorenšek M., Mozetič M. CF4 Plasma and Silver Functionalized Cotton. Textile Research Journal 2010; 80(20) 2204-2213.

28. Mihailovic D, Saponjic Z., Molina R., Puac N., Jovancic P., Nedeljkovic J., Radetic M. Improved Properties of Oxygen and Argon RF Plasma-Activated Polyester Fabrics Loaded with TiO2 Nanoparticles. ACS Applied Materials and Interfaces 2010; 2(6) 1700-1706.

29. Tao D., Feng Q., Gao D., Pan L., Wei Q., Wang D. Surface Functionalization of Polypropylene Nonwovens by Metallic Deposition. Journal of Applied Polymer Science 2010; 117(3) 1624-1630.

30. Yuranova T., Rincon A.G., Bozzi A., Parra S., Pulgarin C., Albers P., Kiwi J. Antibacterial Textiles Prepared by RF-Plasma and Vacuum-UV Mediated Deposition of Silver. Journal of Photochemistry and Photobiology A-Chemistry 2003; 161 (1) 27-34.

31. Pedrosa P., Chappe J.-M., Fonseca C., Machado A. V., Nobrega J.M., Vaz F. Plasma Surface Modification of Polycarbonate and Poly(propylene) Substrates for Biomedical Electrodes. Plasma Processes and Polymers 2010; 7(8) 676-686.

32. Junkar I., Vesel A., Cvelbar U., Mozetič M., Strnad S. Influence of Oxygen and Nitrogen Plasma Treatment on Polyethylene Terephthalate (PET) Polymers. Vacuum 2009; 84 (1) 83-85.

33. Borcia C., Dumitrascu N. Adhesion Properties of Polyamide-6 Fibres Treated by Dielectric Barrier Discharge. Surface and Coatings Technology 2006; 201, 1117-1123.

34. Qi K., Xin J.H., Daoud W.A., Mak C.L. Functionalizing Polyester Fiber with a Self-cleaning Property using Anatase TiO2 and Low-Temperature Plasma Treatment. International Journal of Applied Ceramic Technology 2007; 4(6) 554-563.

35. Mejia M.I., Marin J.M., Restrepo G., Pulgarin C., Mielczarski E., Mielczarski J., Arroyo Y., Lavanchy J.-C., Kiwi J. Self-cleaning Modified TiO2-cotton Pretreated by UVC-light (185 nm) and RF-plasma in Vacuum and also under Atmospheric Pressure. Applied Catalysis B-environmental 2009; 91(1-2) 481-488.

36. Szymanowski H., Sobczyk A., Gazicki-Lipman M., Jakubowski W., Klimek L. Plasma enhanced CVD deposition of titanium oxide for biomedical applications. Surface and coatings technology, 2005; 200(1-4) 1036-1040.

37. Ilic V., Saponjic Z., Vodnik V., Molina R., Dimitrijevic S., Jovancic P., Nedeljkovic J., Radetic M. Antifungal Efficiency of Corona Pretreated Polyester and Polyamide Fabrics Loaded with Ag Nanoparticles. Journal of Materials Science 2009; 44(15) 3983-3990.

38. Gorenšek M., Recelj P. Reactive Dyes and Nano-silver on PA6 Micro Knitted Goods. Textile Research Journal 2009; 79(2) 138-146.

39. Kvitek L., Panacek A., Soukupova J., Kolar M., Vecerova R., Prucek R., Holecova M., Zboril R. Effect of Surfactants and Polymers on Stability and Antibacterial Activity of Silver Nanoparticles (NPs). Journal of Physical Chemistry C 2008; 112(15) 5825-5834.

40. Tomšič B., Simončič B., Cvijn D., Orel B., Zorko M., Simončič A. Elementary Nano Sized Silver as Antibacterial Agent. Tekstilec 2008; 51(7-9) 199–215.

41. Vertelov G., Krutyakov Y., Efremenkova O., Olenin A., Lisichkin G. A Versatile Synthesis of Highly Bactericidal Myramistin Stabilized Silver Nanoparticles. Nanotechnology 2008; 19(35) art.no. 355707.

42. Choi O., Deng K., Kim N., Ross L., Surampalli R., Hu Z. The Inhibitory Effects of Silver Nanoparticles, Silver Ions and Silver Chloride Colloids on Microbial Growth. Water Research 2008; 42(12) 3066-3074.

43. Klemenčič D., Simončič B., Tomšič B., Orel B. Biodegradation of Silver Functionalised Cellulose Fibers. Carbohydrate Polymers 2010; 80, 426-435.

44. Tomšič B., Simončič B., Žerjav M., Simončič A. A Low Nutrition Medium Improves the Determination of Fungicidal Activity of AgCl on Cellulose Fibres. Tekstilec 2008; 51(7-9) 231-241.

45. Tomšič B., Simončič B., Orel B., Žerjav M., Schroers H.-J., Simončič A., Samardžija Z. Antimicrobial Activity of AgCl Embedded in a Silica Matrix on Cotton Fabric. Carbohydrate Polymers 2009; 75(4) 618-626.

46. Gavriliu S., Lungu M., Enescu E., Gavriliu L. Composite Nanopowder for Antibacterial Textiles. Industria Textila 2010; 61(2) 86-90.

47. Yuranova T., Rincon A. G., Pulgarin C., Laub D., Xantopoulos N., Mathieu H.-J., Kiwi J. Performance and Characterization of Ag-cotton and Ag/TiO2 Loaded Textiles during the Abatement of E. coli. Journal of Photochemistry and Photobiology A-chemistry 2006; 181(2-3) 363-369.

48. Rimai D.S., Quesnel D.J., Busnaina A.A. The Adhesion of Dry Particles in the Nanometer to Micrometer-size Range. Colloids and Surfaces A: Physicochemical and Engineering Aspects 2000; 165(1-3), 3-10.

49. Dastjerdi R., Mojtahedi M.R.M., Shoshtari A.M. Investigating the Effect of Various Blend Ratios of Prepared Masterbatch Containing Ag/TiO2 Nanocomposite on the Properties of Bioactive Continuous Filament Yarns. Fibers and Polymers 2008; 9(6) 727-734.

50. Dubas, S.T., Kumlangdudsana, P., Potiyaraj, P. Layer-by-layer Deposition of Antimicrobial Silver Nanoparticles on Textile Fibers. Colloids and Surfaces A: Physicochemical and Engineering Aspects 2006; 289(1-3) 105-109.

51. Hadad L., Perkas N., Gofer Y., Calderon-Moreno J., Ghule A., Gedanken A. Sonochemical Deposition of Silver Nanoparticles on Wool Fibers. Journal of Applied Polymer Science 2007; 104(3) 1732-1737.

52. He J.H., Kunitake T., Nakao A. Facile in situ Synthesis of Noble Metal Nanoparticles in Porous Cellulose Fibers. Chemistry of Materials 2003; 15(23) 4401-4406.

53. Hegemann D., Hossain M.M., Balazs D.J. Nanostructured Plasma Coatings to Obtain Multifunctional Textile Surfaces. Progress in Organic Coatings 2007; 58(2-3) 237-240.

54. Moazami A., Montazer M., Rashidi A., Rahimi M.K. Antibacterial Properties of Raw and Degummed Silk with Nanosilver in Various Conditions. Journal of Applied Polymer Science 2010; 118(1) 253-258.

55. Gorensek M., Recelj P. Nanosilver Functionalized Cotton Fabric. Textile Research Journal 2007; 77(3) 138-141.

56. Jiang S.X., Qin W.F., Zhang L. Surface Functionalization of Nanostructured Silver-Coated Polyester Fabric by Magnetron Sputtering. Surface and Coatings Technology 2010; 204(21-22) 3662-3667.

57. Malinsky M.D., Kelly K.L., Schatz G.C., Van Duyne R.P. Chain length dependence and sensing capabilities of the localized surface plasmon resonance of silver nanoparticles chemically modified with alkanethiol self-assembled monolayers, J. Am. Chem. Soc., 123 (2001), pp. 1471–1482.

58. Bosbach J., Hendrich C., Stietz F., Vartanyan T., Trager F. Ultrafast Dephasing of Surface Plasmon Excitation in Silver Nanoparticles: Influence of Particle Size, Shape, and Chemical Surrounding. Physical Review Letters 2002; 89(25) art.no. 257404.

59. Mock J.J., Barbic M., Smith D.R., Schultz D.A., Schultz S. Shape Effects in Plasmon Resonance of Individual Colloidal Silver Nanoparticles. Journal of Chemical Physics 2002; 116(15) 6755-6759.

60. Ilic V., Saponjic Z., Vodnik V., Potkonjak B., Jovancic P., Nedeljkovic J., Radetic M. The Influence of Silver Content on Antimicrobial Activity and Color of Cotton Fabrics Functionalized with Ag Nanoparticles. Carbohydrate Polymers 2009; 78 (3) 564-569.

61. Carneiro N., Souto A.P., Silva E., Marimba A., Tena A., Ferreira H., Magalhaes V. Dyeability of Corona-Treated Fabrics. Coloration Technology 2001; 117, 298-302.

62. Navarro, A., Bautista, L. Surface modification and characterization in cotton fabric bleaching. Proceedings of 5th world textile conference AUTEX 2005. Maribor; 2005.

63. Gorjanc M., Jovančić P., Bukošek V., Gorenšek M. Study of adsorption of nano silver on cotton pretreated with plasma. Proceedings of the 9th Autex, Izmir, Turkey; 2009.

64. Gorjanc M., Jovančić P., Gorenšek M. Nano silver on cotton. In: Mihailović D., Kobe S., Remškar M., Jamnik J., Čopič M., Drobne D. (eds.). Hot nano topics 2008 : incorporating SLONANO 2008, 3 overlapping workshops on current hot subjects in nanoscience, 23-30 May, Portorož, Slovenia : abstract book. Ljubljana; 2008.

65. Gorjanc M., Jovančić P., Kovač J., Bukošek V., Gorenšek M. Plasma treatment of cotton: correlation between dyeability of cotton with reactive dye and chemical composition of surfaces. In: XXII IFATCC International congress (International Federation of the Association of Textile Chemists and colorists), Lago Magiore (Italy), May 5-7th, 2010. From textile chemistry to fashion : multifunctionality, sustainability, competitivity : proceedings of the XXII IFATCC International congress, Stresa, Lago Magiore. Milano: Associazione Italiana di Chimica Tessile e Coloristica: = AICTC Italia; 2010.

66. Gorjanc M., Mozetič M., Bukošek V., Gorenšek M. Impact of low-pressure plasma on dyeability of cotton. In: Dragčević Z. (ed.) 4th International Textile, Clothing & Design Conference [also] ITC&DC, October 5th to October 8th, 2008, Dubrovnik, Croatia. Magic world of textiles: book of proceedings. Zagreb: Faculty of Textile Technology, University of Zagreb; 2008.

67. Gorjanc M., Mozetič M., Gorenšek M. Low-pressure plasma for pretreatment of cotton fabric for better adhesion of nanosilver. Tekstilec 2009; 52(10-12) 263-269.

68. Ames Goldsmith - Silver nano powders. Data sheet, 2005. http://www.amesgoldsmith.com/products/silver-nano-powders.asp/ (accessed 25 February 2011).

69. Gorjanc M. Plasma Modification of Cotton Surfaces for a Deposition of Nanosilver. PhD thesis. University of Ljubljana, Faculty of Natural Sciences and Engineering; 2011.

70. The ultimate hygenic finish for textiles. Rudolph Chemie, Rudolph Info 5/07; 2007.

71. Gorjanc M., Kovač F., Gorenšek M. The Influence of Vat Dyeing on the Adsorption of Synthesized Colloidal Silver onto Cotton Fabrics. Textile Research Journal 2012; 82(1) 62-69.

72. Gorjanc M., Gorenšek M. Influence of Dyeing Cotton with Reactive Dye on Adsorption of Silver. Tekstilec 2011; 54(10-12) 228-237.

73. Doering R., Nishi Y. Handbook of Semiconductor Manufacturing Technology. Boca Raton : CRC Press; 2000.

74. Johansson L.S., Campbell J.M., Stenius P., Pere J., Buchert T. Surface Characterization of Cellulosic Materials with XPS. Abstracts of Papers of the American Chemical Society 2001; 221 (Part I) U180-U180.

75. Gorjanc M., Mozetič M., Jazbec K., Gorenšek M. Influence of ICRF water vapour plasma on dyeability and UPF of cotton fabrics. In: Simončič B., Hladnik A., Pavko-Čuden A., Ahtik J., Luštek Preskar B., Demšar A., Urbas R. (eds.). 41st International Symposium on Novelties in Textiles and 5th International Symposium on Novelties in Graphics and 45th International Congress IFKT, Ljubljana, Slovenia, 27-29 May 2010. Symposium proceedings. Ljubljana: Faculty of Natural Sciences and Engineering, Department of Textiles; 2010.

76. Tsafack M.J., Levalois-Grutzmacher J. Towards Multifunctional Surfaces using the Plasma-Induced Graftpolymerization (PIGP) Process: Flame and Waterproof Cotton Textiles. Surface and Coatings Technology 2007; 201(12): 5789-5795.

77. McCord M.G., Hwang Y.J., Qui Y., Hughes L.K. Bourham M.A. Surface Analysis of Cotton Fabrics Fluorinated in Radio-Frequency Plasma. Journal of Applied Polymer Science 2003; 88(8), 2038-2047.

78. Cvelbar U., Krstulovic N., Milosevic S., Mozetic M. Inductively Coupled RF Oxygen Plasma Characterization by Optical Emission Spectroscopy. Vacuum 2007; 82(2) 224-227.

79. Kurihara M., Petrovic Z., Makabe T. Transport Coefficients and Scattering Cross-Sections for Plasma Modelling in CF4-Ar Mixtures: a Swarm Analysis. Journal of Physics D-Applied Physics 2000; 33(17); 2146-2153.

80. Flamm D.L. Mechanisms of Radical Production in CF3Cl, CF3Br, and Related Plasma Etching Gases: The role of Added Oxidants. Plasma Chemistry and Plasma Processing 1981; 1(1) 37-52.

81. d'Agosiono R. Plasma Deposition, Treatment, and Etching of Polymers. San Diego: Academic Press Inc; 1990.

82. Poll, H.U., Schladitz, U., Schreiter, S. Penetration of Plasma Effects into Textile Structures. Surface and Coatings Technology 2001; 142, 489-493.

83. Hwang Y.J., McCord M.G., An J.S., Kang B.C., Park S.W. Effects of Helium Atmospheric Pressure Plasma Treatment on Low-Stress Mechanical Properties of Polypropylene Nonwoven Fabrics. Textile Research Journal 2005; 78(11) 771-778.

84. Yip J., Chan K., Sin K.M., Lau K.S. Study of Physico-Chemical Surface Treatments on Dyeing Properties of Polyamides. Part 1: Effect of Tetrafluoromethane Low Temperature Plasma. Coloration Technology 2002; 118(1) 26-30.

85. Sun D., Stylios G.K. Investigating the Plasma Modification of Natural Fiber Fabrics - the Effect on Fabric Surface and Mechanical Properties. Textile Research Journal 2005; 75(9) 639-644.

86. Owens F.J., Poole Jr., C.P., editors. The physics and chemistry of nanosolids. New Jersey: Wiley & Sons, Inc.; 2008.

87. Kawabata A., Taylor J.A. Effect of Reactive Dyes upon the Uptake and Antibacterial Action of Poly(Hexamethylene Biguanide) on Cotton. Part 1: Effect of Bis(Monochlorotriazinyl) Dyes. Coloration Technology 2004; 120(5) 213-219.

88. Preston C., editor. The Dyeing of Cellulosic Fibers. London: Dyers' Company Publications Trust; 1986.

89. Gorenšek M. Dyes and textile colour. In: Jeler S. (ed.) Textile and colour. Firenze: Edizioni Tassinari; 2004. P5-19.

90. Shore J. Cellulosics Dyeing. Bradford : Society of Dyers and Colourists; 1995.

91. Nair, G.P., Shah, R.C. Sodium Borohydride in Vat Dyeing. Textile Research Journal 1970; 40(4) 303-312.

92. Klemenčič D., Simončič B., Tomšič B., Orel B. Biodegradation of Silver Functionalized Cellulose Fibres. Carbohydrate Polymers 2010; 80, 426-435.

CHAPTER 3

Characterization of Natural Dyes and Traditional Korean Silk Fabric by Surface Analytical Techniques

Jihye Lee [1], Min Hwa Kang [1,2], Kang-Bong Lee [1] and Yeonhee Lee [1,*]

[1] Advanced Analysis Center, Korea Institute of Science and Technology, Seoul 136-791, Korea

[2] Department of Nano and Bio Surface Science, University of Science and Technology, Daejeon 305-350, Korea

ABSTRACT

Time-of-flight secondary ion mass spectrometry (TOF-SIMS) and X-ray photoelectron spectroscopy (XPS) are well established surface techniques that provide both elemental and organic information from several monolayers of a sample surface, while also allowing depth profiling or image mapping to be carried out. The static TOF-SIMS with improved performances has expanded the application of TOF-SIMS to the study of a variety of organic, polymeric and biological materials. In this work, TOF-SIMS, XPS and Fourier Transform

Infrared (FTIR) measurements were used to characterize commercial natural dyes and traditional silk fabric dyed with plant extracts dyes avoiding the time-consuming and destructive extraction procedures necessary for the spectrophotometric and chromatographic methods previously used. Silk textiles dyed with plant extracts were then analyzed for chemical and functional group identification of their dye components and mordants. TOF-SIMS spectra for the dyed silk fabric showed element ions from metallic mordants, specific fragment ions and molecular ions from plant-extracted dyes. The results of TOF-SIMS, XPS and FTIR are very useful as a reference database for comparison with data about traditional Korean silk fabric and to provide an understanding of traditional dyeing materials. Therefore, this study shows that surface techniques are useful for micro-destructive analysis of plant-extracted dyes and Korean dyed silk fabric.

KEYWORDS

natural dyes; traditional silk fabric; mordants; TOF-SIMS; XPS; FTIR; surface techniques

1. INTRODUCTION

Organic dyes are very significant materials in archaeological remains. Investigation of the coloring materials used on ancient textiles provides a historical window to understanding the surrounding environment, the living style and the technological skill in ancient times [1,2,3,4]. However, dyes have complicated chemical composition and can co-exist with other chromophores and contaminants, so that the identification of organic dyes in old textiles is a challenging task. There are four major barriers to characterizing organic dyes: (1) the relatively small quantity of samples available; (2) the low concentration of chromophores in the textiles; (3) the presence of possible degradation products; and (4) the lack of information on the original recipes [5,6,7].

For last two decades, chemical research in the field of the analytical separation and spectrometric identification of organic dyes has been performed and studies have presented a new chromatographic scheme for the identification of certain dyes [8]. The identification of organic dyes in various types of textiles was conducted using chromatographic techniques, such as liquid chromatography mass spectrometry, gas chromatography mass spectrometry, high performance liquid chromatography and thin-layer chromatography [9,10,11,12]. The extraction of dyes from textiles is a critical process in this type of analytical techniques due to the small amount of sample available. Using HPLC analysis, a successful new mild extraction method was established for the identification of dyes in historical remains [13,14].

Recently, researchers have focused great attention on analytical techniques based on direct mass spectrometry, such as direct exposure mass spectrometry (DE-MS), laser desorption/ionization mass spectrometry (LDI-MS), matrix assisted laser desorption/ ionization mass spectrometry (MALDI-MS) and time-of-flight secondary ion mass spectrometry (TOF-SIMS) [15,16,17,18]. Compared to wet chemical preparation, these techniques involve minimum sample requirement, no sample pretreatment, less contamination and reduced samples loss. Furthermore, the analysis times are very short and the sensitivities are suitable for identifying organic dyes in textiles. Lee *et al.* used TOF-SIMS to study natural dyes and ancient textiles from Korea [17]. Indigo was successfully identified by TOF-SIMS for two ancient textiles from 16th and 17th-century Korea.

In this study, comparing to the conventional FTIR technique, to avoid extraction and to use minimum sample amount, surface analytical techniques, such as TOF-SIMS and X-ray photoelectron spectroscopy (XPS), were used to detect natural dyes in silk fabric. TOF-SIMS has been found to be an ideally suited analytical technique to characterize surface compositions and structures, because it provides elemental and molecular information about the surface layers with a high degree of sensitivity, mass resolution and special resolution [19]. XPS occupies a unique position in the field of archaeological analysis, because it permits examination of an extensive array of elements from the surfaces of artifacts and because it offers sensitivity to elemental oxidation states [20]. FTIR spectroscopy has been applied to the characterization of specific dyes in textiles; however, signals from the matrix severely interfere with the identification of the coloring materials [21]. Therefore, a complementary analysis of various natural dyes and silk fabric dyed with plant extracts was performed by TOF-SIMS, XPS and FTIR.

2. EXPERIMENTAL SECTION

2.1. Materials

The commercial natural dyes, such as curcumin, indigo and quercetin, were purchased from the Tokyo Chemical Industry (TCI) Co., LTD. Tokyo, Japan. Both brazilin and shikonin were supplied by the Sigma-Aldrich Chemical Company, Inc., St. Louis, MO, USA. Natural dyes were also extracted from four plants. In this case, turmeric and sappanwood plants were imported from India and the Philippines, whilst gromwell and indigo plants were all readily available in Korea.

2.1.1. Dyeing Using Plant Extracts

The majority of the natural plant dyes were extracted in water either by soaking at room temperature or by heating. Turmeric (20 g) was placed in 1 L of methanol for 7 days. Silk fabric (10 cm × 10 cm) was soaked in 2.5 mL of turmeric extract diluted with 300 mL of water at 40 °C for 60 min. Indigo plant was soaked in water for 4 days. Then, $Ca(OH)_2$ was added to the solution of 250 mL, and the resulting precipitate was filtered and dried. Silk fabric of 10 cm × 10 cm has been dyed in 1.7 L of water mixed with 100 g of precipitated dye. For zinc mordanting, silk fabric was treated in 1.7 L of water containing 100 g of dye, 3.3 g of zinc powder and 7 g of sodium hydroxide at room temperature for 60 min. Sappanwood (200 g) was soaked in 2 L of water and solution was heated. Silk fabric (10 cm × 10 cm) was dyed in dye solution for 30 min. After dyeing, the dyed fabric was washed in running water and dried at room temperature. Silk fabric was treated with different metal salts. In the case of Al mordant, silk fabric was treated in 400 mL of dye solution containing 2 g of alum at 40 °C for 30 min. For Fe mordant, silk fabric was dyed in 200 mL of dye solution for 30 min and then treated in 200 mL of water containing 3 g of iron sulfate (II) for 15 min. Sn mordanting was carried out in 200 mL of water with 0.08 g of tin chloride (IV) for 15 min and dyed in dye solution for 30 min. In the case of Cu mordant, silk fabric was dyed in dye solution at 40 °C for 30 min and washed. After pre-dyeing, fabric was treated in copper sulfate for 15 min and rinsed with water. Gromwell (50 g) was soaked in water at 40 °C for 3 h and peeled slightly. Then, colorant was extracted kneading macerated gromwell in 1 L of water at 60 °C for 20 min. After filtering the solution, silk fabric (10 cm × 10 cm) was dyed in dye solution at 60 °C for 20 min two times. Dyed silk fabric was rinsed with water and dried at room temperature.

2.1.2. Dyeing Using Commercial Dyes

Silk fabric (10 cm × 10 cm) was dyed in 300 mL of water containing 1 g of curcumin powder at 60 °C for 30 min and dyeing process was repeated two times. In order to use indigo for dyeing fabrics, it must be transformed to the soluble form of indigo, known as leuco-indigo. Five grams of sodium hydroxide was dissolved in 200 mL of distilled water and added 5 g of sodium hydrosulfite. After removing insoluble material, silk fabric (10 cm × 10 cm) was dyed in 55 °C for 30 min and rinsed in cold water. Silk fabric (10 cm × 10 cm) was dyed in 300 mL of water containing 1 g of brazilin, shikonin or quercetin powder at 40 °C for 30 min and dyeing process was repeated two times. A detailed description of dyeing methods is presented elsewhere [22]. Table 1 shows the empirical formula, molecular weight and chemical structures of the major components of plant dyes selected for this work. Table 2 shows various colors and colorimetric parameter of textiles dyed with commercial dyes and natural dyes extracted from plants.

2.2. Instrumental Evaluation

Time-of-Flight Secondary Ion Mass Spectrometry. Positive and negative ion TOF-SIMS analyses were carried out on an ION-TOF GmbH (Münster, Germany). A TOF-SIMS 5 system was equipped with a Bi^+ primary ion beam source. A Bi^+liquid metal ion source (LMIS) was employed for mass data acquisition. Such cluster ions provide better secondary ion yields than those available from the more conventional Ga^+ source. The analysis source was a pulsed 25-keV Bi^+ beam that bombarded the surface at an incident angle of 45° to the surface normal. The target current was maintained at a pulsed current of 1 pA with a raster size of 100 μm × 100 μm. A cycle time of 100 μs was employed for mass data acquisition. These conditions resulted in Bi^+ ion doses that were well below the static SIMS limit of 10^{13} ions/cm^2.

Table 1. Details of the natural dyes extracted from plants.

Major dye component MW Color	**Name of plant (English)**	**Chemical structure**
Curcumin $C_{21}H_{20}O_6$ 368.38 Yellow	*Curcuma longa L.* (Turmeric)	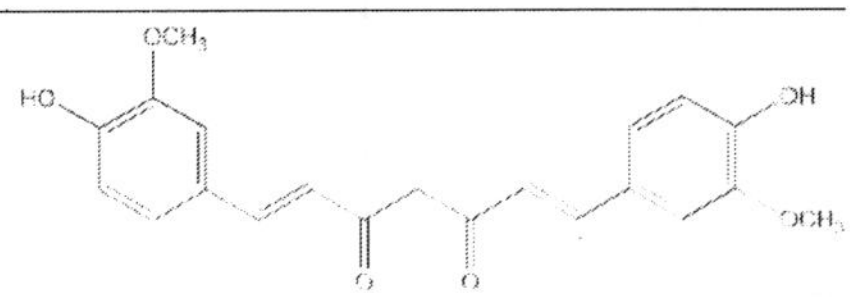
Indigo $C_{16}H_{10}N_2O_2$ 262.26 Blue	*Indigofera tinctoria* (Indigo)	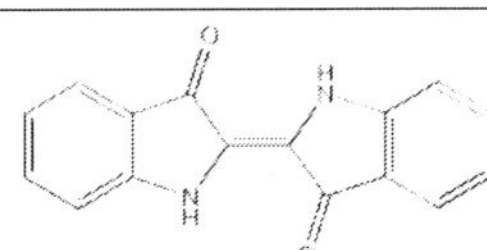
Brazilin $C_{16}H_{14}O_5$ 286.27 Red	*Caesalpinia sappan L.* (Sappanwood)	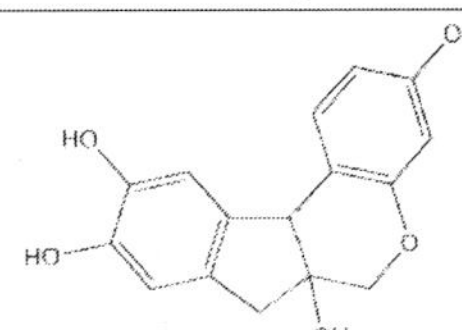
Shikonin $C_{16}H_{16}O_5$ 288.29 Purple	*Lithospermum erythrorhizon S. et Z* (Gromwell)	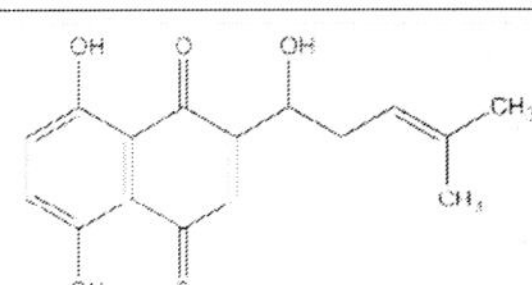
Quercetin $C_{15}H_{10}O_7$ 302.24 Yellow	*Sophora japonica L.* (Japanese pagoda tree)	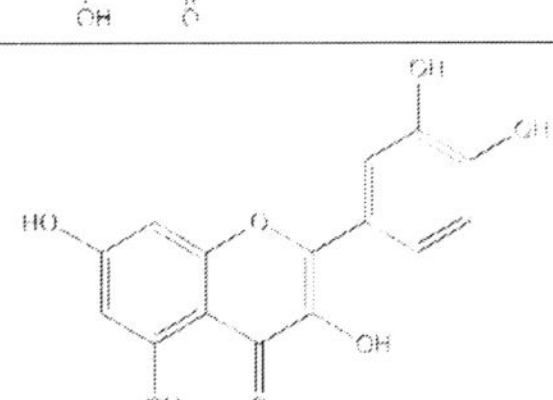

Table 2. Various colors and colorimetric data of silk fabric dyed with commercial dyes, plant extracts dyes and plant extracts dyes with mordants. (L*,a*,b* is CIELAB color scale)

Commercial dyes			Natural dyes from plant extracts			Plant extracts dyes + Mordants		
	L*	87.30		L*	85.83		L*	31.97
	a*	−12.88		a*	−8.09		a*	0.39
Curcumin	b*	69.33	Turmeric	b*	48.63	Indigo (Zn)	b*	−18.11
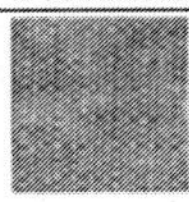	L*	51.82		L*	49.78		L*	37.20
	a*	−3.38		a*	−7.27		a*	42.10
Indigo	b*	−19.96	Indigo	b*	−20.55	Sappanwood (Al)	b*	22.38
	L*	69.49	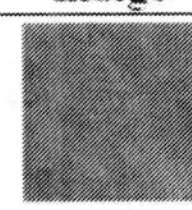	L*	58.92	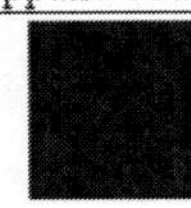	L*	19.81
	a*	2.59		a*	21.05		a*	2.49
Brazilin	b*	29.67	Sappanwood	b*	42.79	Sappanwood (Fe)	b*	1.03
	L*	67.77		L*	−10.30		L*	61.51
	a*	18.67		a*	19.66		a*	28.71
Shikonin	b*	4.46	Gromwell	b*	−18.12	Sappanwood (Sn)	b*	27.81
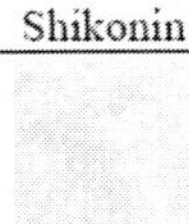	L*	76.62	–				L*	34.77
	a*	−3.52					a*	21.35
Quercetin	b*	13.40				Sappanwood (Cu)	b*	11.95

Both positive and negative secondary ions were extracted from the sample and accelerated using an ion lens operating at 2 kV into a time-of-flight (TOF) mass analyzer. The TOF uses a reflectron to reduce line widths and has a total flight path of 2 m. The secondary ion beam is focused on the microchannel plate with a post-acceleration energy of 20 kV. A low-energy electron flood gun was used for charge compensation of all silk textile samples.

X-ray Photoelectron Spectroscopy. X-ray photoelectron spectroscopy (XPS) was used to examine the chemical structure on the surface of natural dyes on silk fabric using a PHI-5000 VersaProbe instrument (ULVAC-PHI, Kanagawa, Japan) with a base pressure of 6.7 × 10^{-8} Pa. The dyed silk textiles were analyzed by measuring the C1s, N1s and O1s and the metal element spectra in the XPS. The XPS analysis was conducted with monochromatic Al Kα radiation (1486.6 eV) at a power of 25 W. The instrument was calibrated to the $4f_{7/2}$ peak of gold at E_b = 84.03 eV with a full width at half maximum (FWHM) of 0.71 eV and to the copper $2p_{3/2}$ peak at E_b = 932.66 eV with a FWHM of 0.91 eV. Charging of the dyed silk textile, as a result of photoemission, was corrected by setting the energy of the main hydrocarbon component of the C1s spectrum to 284.6 eV for dyed textiles. A pass energy of 23.5 eV was utilized for all narrow spectra. The spot size of the X-ray beam was 100 μm × 100 μm in each case. The sample was exposed to X-rays for several minutes to minimize the X-ray beam damage.

Fourier Transform Infrared Spectroscopy. Attenuated Total Reflection Fourier Transform Infrared (ATR-FTIR) measurements were carried out with an Infinity Gold FT-IR Series (ThermoMattson, Waltham, MA, USA) system equipped with an IR microscope (SpectraTech, Inspect IR plus, Franklin Lakes, NJ, USA) with a mercury cadmium telluride detector. One hundred twenty eight scans were collected with a resolution of 4 cm^{-1}, and the frequency range was measured between 600 and 4000 cm^{-1} of the mid-IR region. Dyed silk fabric were mounted in the sample holder and no additional sample preparation was done. Spectrum of the undyed silk textile was recorded and this served as the blank throughout the experiment. Several spectra of the undyed silk textile were done to check for the reproducibility.

Color measurements. Color measurements of the dyed silk fabric were performed directly using a spectrophotometer (Minolta Chromameter, CM-5).

3. RESULTS AND DISCUSSION

3.1. FT-IR Measurement

Table 1 shows the empirical formulas, molecular weights and chemical structures of the major components of the plant dyes selected for this work; these include curcumin, indigo, brazilin, shikonin and quercetin. To establish the FTIR technique for dye identification, commercial dyes and the dyed silk fabric were prepared. Figure 1 compares the FTIR spectra of curcumin, indigo, brazilin, shikonin and quercetin-dyed silk fabric obtained by subtraction of undyed silk with one obtained from the commercial dyes. Undyed silk textile was used to obtain the reference spectrum. No dye component could be detected in the FTIR spectra unless spectral subtractions were conducted. After spectral subtraction, it was easily possible to identify the dye component in all of the dyed silk fabric.

The spectrum of indigo is shown with a specific peak at 2360 cm^{-1} due to the N–H stretching, a peak at 1628 cm^{-1} that is because of C=O stretching and a peak at 1392 cm^{-1} that is because of N–H bending. Other specific peaks are the aromatic ring C=C vibration at 1461, 1483 and 1585 cm^{-1}.

Minerals readily available from the environment were used in combination with natural dyes to provide a strong bond between the textile and the dye. Color brightness and fastness are improved by the use of mordants in the dyeing process. However, the mordant had no apparent effect on the FTIR spectra. The FTIR spectra of the indigo extract-dyed and zinc-mordanted silk fabric that underwent spectral subtraction of undyed silk from indigo-dyed silk matched those of silk fabric dyed with commercial indigo, with specific peaks at 1628, 1124 and 1068 cm^{-1} (Figure 2). These results, which are expected, because indigo is the main coloring component in indigo plant, support the accuracy of the FTIR technique. However, it was not possible to distinguish individual dye components in silk fabric dyed with the other natural extracts using the FTIR technique. In this study, we prepared dyed silk fabric using the traditional dying process; the amounts of dyes used in this procedure may be below the detection limit of FTIR.

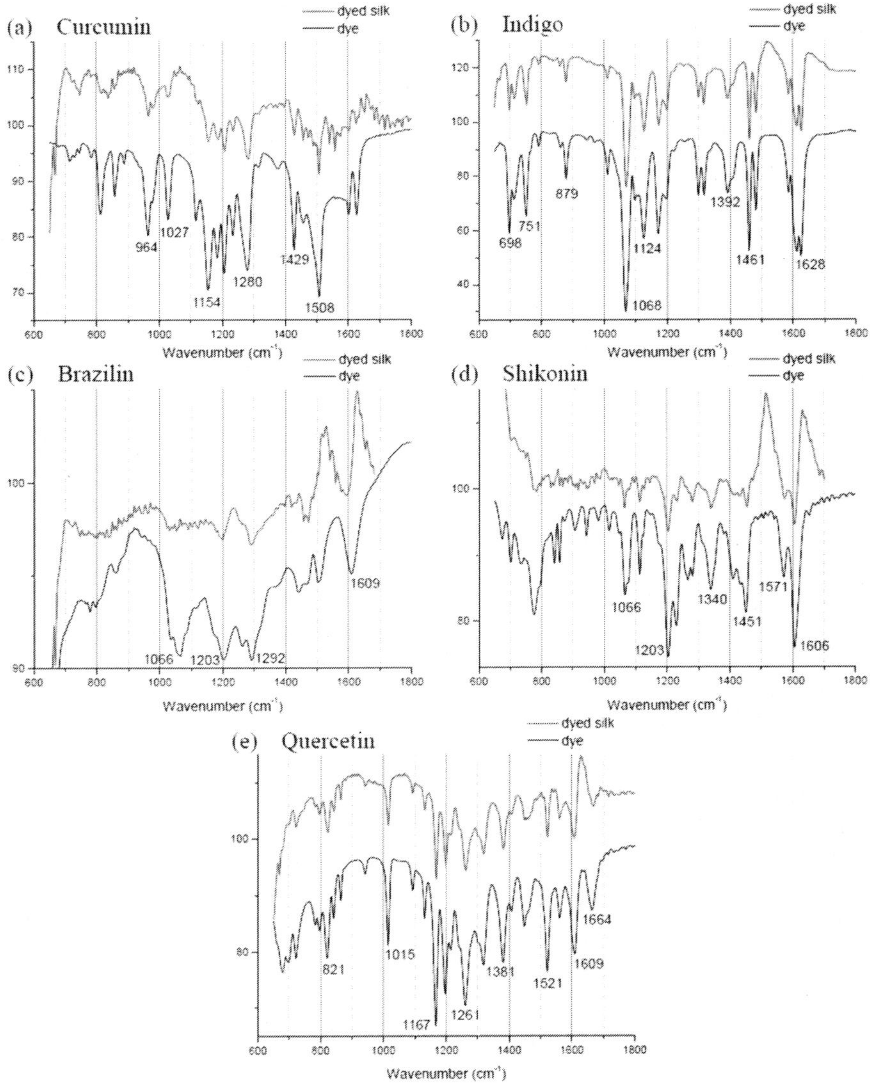

Figure 1. Fourier transform infrared spectra of five commercial dyes and the dyed silk fabric between 600 and 1800 cm^{-1}: (**a**) curcumin; (**b**) indigo; (**c**) brazilin; (**d**) shikonin; and (**e**) quercetin.

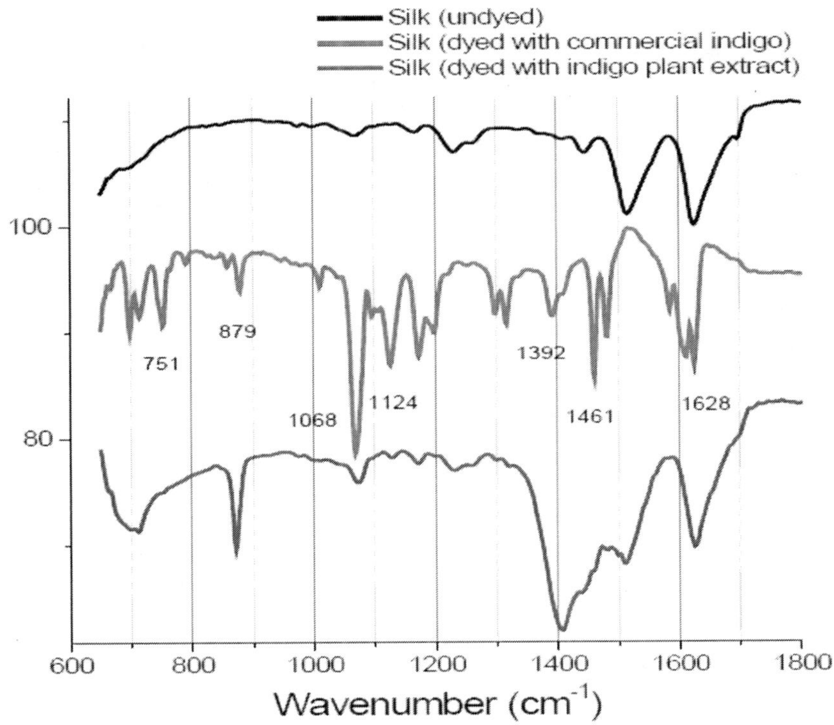

Figure 2. Fourier transform infrared spectra of undyed silk textile with commercial indigo and silk textile dyed with indigo plant extract. The characteristic peaks allow easy identification. The spectra are vertically shifted for sake of clarity.

3.2. XPS Measurement

XPS techniques have rarely been applied to the identification of specific dyes in textiles, because signals due to the textile heavily interfere with the identification of coloring materials. The XPS survey spectra for dyed silk fabric were obtained to determine chemical compositions of silk textiles. C1s and O1s peaks are the main signals in the survey spectra of the silk fabric dyed with curcumin, brazilin, shikonin and quercetin, while C1s, O1s and N1s peaks are the main signals for those dyed with indigo. As can be seen in Figure 3, the survey spectra of sappanwood-dyed silk fabric reveal the presence of aluminum, iron, tin and copper due to the addition of mordants during the dyeing process. The calculated atomic concentrations obtained by XPS for the sappanwood-dyed silk fabric are listed in Table 3. As expected, the surface compositions of these silk textiles are similar. The discrepancy of carbon and oxygen concentrations from the theoretical values was probably caused by hydrocarbon and remaining vapor contamination, the textile surface and the surface degradation induced by X-ray. The amounts of mordants are rather low, so their peaks are not intense in the XPS spectra. However, quantitative results for various mordant elements in dyed silk fabric were obtained by the XPS technique.

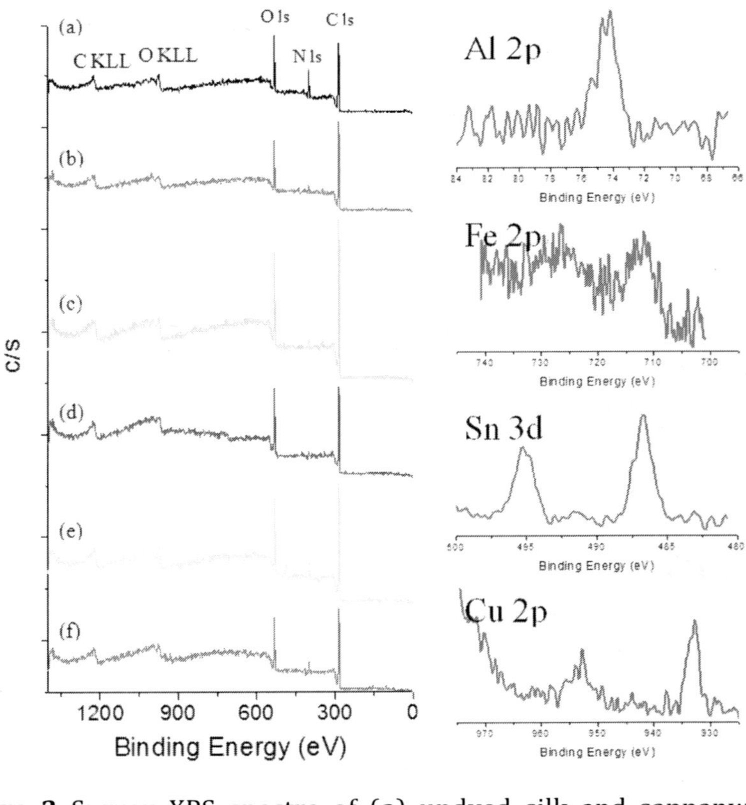

Figure 3. Survey XPS spectra of (**a**) undyed silk and sappanwood-dyed silk textiles; (**b**) without mordant and with various mordants: (**c**) $KAl(SO_4)_2$; (**d**) $FeSO_4$; (**e**) $SnCl_4$; and (**f**) $CuSO_4$.

Table 3. XPS Surface compositions of silk fabric dyed with sappanwood extracts and different mordants.

Sample/Mordant	Elements composition (%)						
	C (%)	O (%)	N (%)	Al (%)	Fe (%)	Sn (%)	Cu (%)
Sappanwood	77.05	18.28	4.66	–	–	–	–
Sappanwood/$K.Al(SO_4)_2$	76.28	21.14	1.53	1.04	–	–	–
Sappanwood/$FeSO_4$	74.21	22.77	2.17	–	0.85	–	–
Sappanwood/$SnCl_4$	73.18	20.34	6.12	–	–	0.36	–
Sappanwood/$CuSO_4$	73.66	20.62	4.63	–	–	–	1.09

To offset the surface charging phenomena, all binding energies in the C1s core level spectra were referenced to the C–C peak centered at 284.6 eV. The peak deconvolution of the XPS result was performed by Gaussian-Lorentz mixture function using MULTI-PAK v9.1 software. Figure 4 illustrates C1s peaks of dyed silk fabric resolved using curve fitting. The XPS results for the silk fabric dyed with turmeric, indigo, sappanwood, gromwell and quercetin are listed in Table 4. The table presents the percentages of each carbon functional group presented in the spectra. The peak fitting was performed with a consideration of parameters, such as binding energy, full width at half maximum and Gaussian character. The fitting parameters for the dye components of the dyed silk textile in the C1s spectrum were determined by analyzing the C1s spectrum measured for a sample of commercial dye. Line-shape analysis using peak deconvolution shows that the C1s spectrum for undyed silk textile contains three distinct peaks at 284.6 (–C–C), 286.2 (–C–OH or –C–N) and 288.0 eV (N–C=O). These peaks may be attributed to the bonds present in the silk and to any residual surface contaminants. After the dyeing process with plant extracts, the intensity of C1s increased and C1s spectra showed peaks of oxygen-containing groups at 286.2 (–C–OH or –C–O–C) and 287.6 eV (–C=O) with different intensities. As shown in the chemical structures ofTable 1, turmeric and quercetin have more oxygen-containing functional groups, like hydroxyl and carboxyl groups, than gromwell. XPS results also indicated that turmeric and quercetin had more C–OH and C=O bonds than gromwell. The XPS technique provided the quantification results for the surface composition of the dyed silk textiles with various plant extracts and mordants.

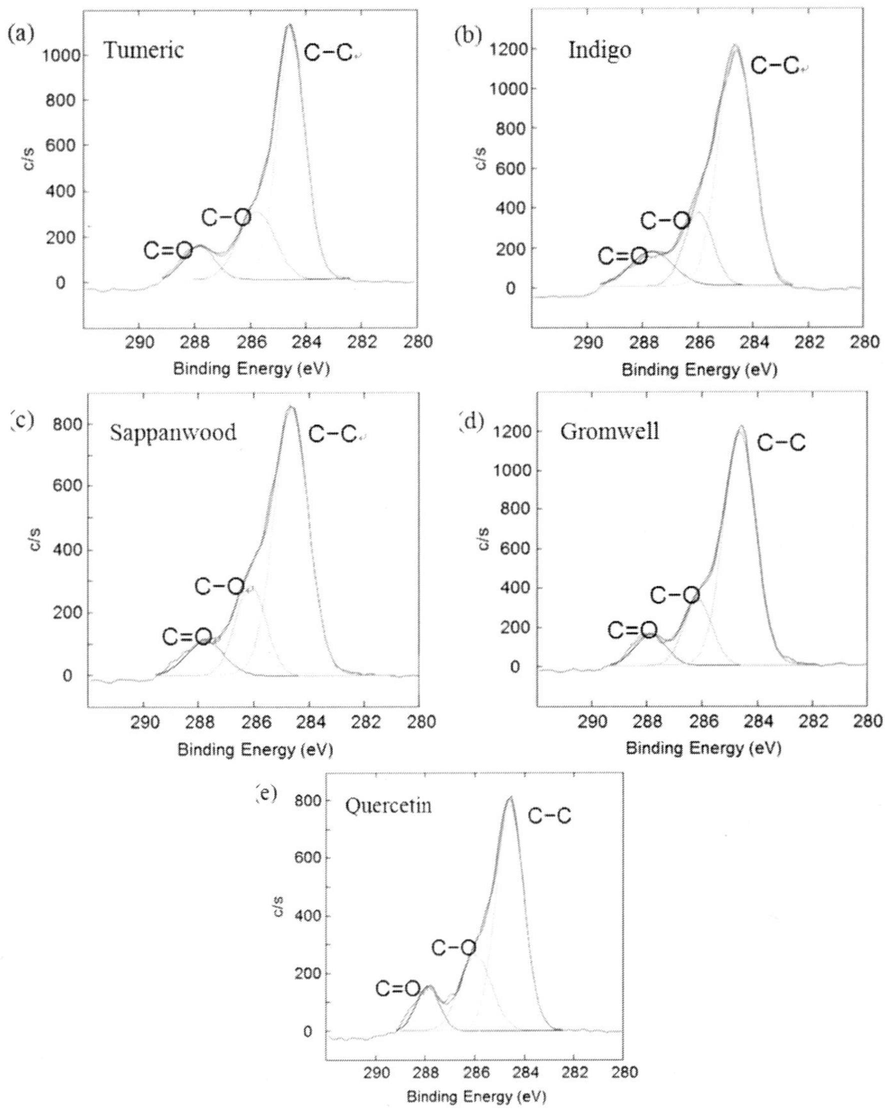

Figure 4. Results of the curve fitting analysis of C1s spectrum measured for silk textiles dyed with plant extracts. (**a**) Turmeric; (**b**) Indigo; (**c**) Sappanwood; (**d**) Gromwell; (**e**) Quercetin.

Table 4. Relative concentration of C1s components for silk fabric dyed with plant extracts.

Sample	Relative chemical bond area Cls (%)		
	C–C (284.6 eV)	C–O C–N (286.2 eV)	C=O (287.7 eV)
Turmeric	67.22	23.28	9.51
Indigo	69.55	18.08	12.37
Sappanwood	69.92	20.32	9.76
Gromwell	73.10	17.73	9.17
Quercetin	65.17	24.26	10.58

3.3. TOF-SIMS Measurement

The commercial dyes used in this work were curcumin, indigo, brazilin, shikonin and quercetin. Five commercial natural dyes were purchased from the manufacturer, and their TOF-SIMS spectra were obtained from thin layer films cast from solutions of dyes in solvent on a silicon substate. The TOF-SIMS technique provides molecular structure and composition information from the uppermost layer of the dye components. To characterize the commercial natural dyes, the positive ion TOF-SIMS mass spectra were acquired at the film surface of the dyes. The positive ion mass spectra of the dyes were acquired using a Bi^+ primary ion beam under static SIMS conditions from a solution-cast dye film exposed to room temperature, as shown in Figure 5.

Interpretation of the spectra will focus on the information they provide for structural characterization. A segment from a typical spectrum of dyes in the mass range of m/z = 0–100 shows that the dominant peaks are $C_2H_3^+$ (m/z 27.023), $C_2H_5^+$(m/z 29.039), $C_3H_5^+$ (m/z 41.039), $C_3H_7^+$ (m/z 43.055), $C_2H_3O^+$ (m/z 43.018), $C_4H_7^+$ (m/z 55.055), $C_3H_3O^+$ (m/z 55.018), $C_4H_9^+$(m/z 57.070) and $C_3H_5O^+$ (m/z 57.034). Na^+ (m/z 22.990) and K^+ (m/z 38.964) ions are also found in the spectra, as well as Si^+(m/z 27.977) from the silicon substrate (not shown in the figure). The other peaks in the lower mass range are due to carbon- and oxygen-containing ions that are found in the positive spectra of nearly all organic materials.

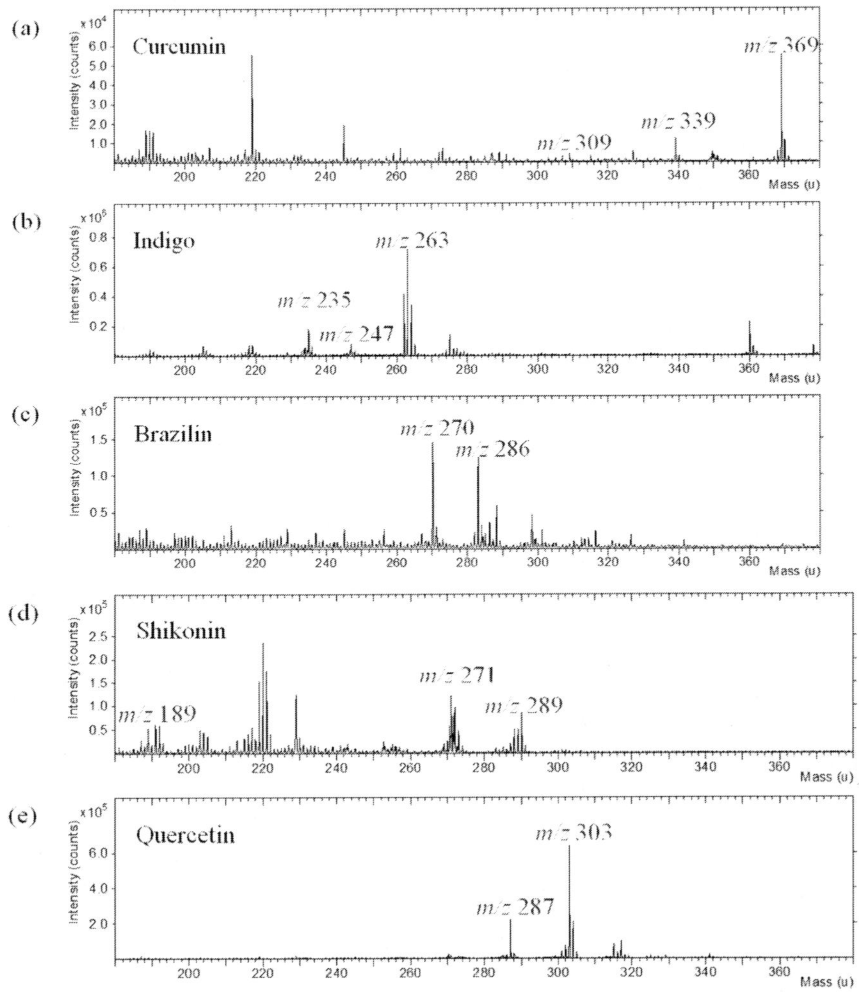

Figure 5. Positive ion TOF-SIMS spectra of commercial natural dyes on the silicon substrate. (**a**) curcumin; (**b**) indigo; (**c**) brazilin; (**d**) shikonin; (**e**) quercetin.

In the higher mass range between m/z 180 and 380, the positive TOF-SIMS spectra of curcumin (Figure 5a) indicate the presence of molecular ions, M+H at m/z 369 and specific fragment ions, *i.e.*, $C_{10}H_9O_3^+$ at m/z 177, $C_{19}H_{17}O_4^+$ at m/z 309 and $C_{20}H_{19}O_5^+$ at m/z 339. The positive ion spectrum of indigo shows characteristic ions at m/z

120 ($C_7H_6NO^+$), 132 ($C_8H_6NO^+$), 235($C_{15}H_{11}N_2O^+$) and 247 ($C_{16}H_{11}N_2O^+$), as well as molecular ion (M+H) at m/z 263. TOF-SIMS spectra of the other dye compounds studied, such as brazilin, shikonin and quercetin, also show molecular ions at m/z 286 (M^+), m/z 289 (M+H) and m/z 303 (M+H) and specific fragment ions in the positive ion spectra.

Silk fabric were dyed with commercial natural dyes. TOF-SIMS analysis of dyed silk textile provided complicated spectra composed of peaks from dyes, silk fabric and contaminants. The TOF-SIMS spectrum of silk textile was also obtained to discriminate the background information from the silk textile. The silk textile gives hydrocarbon ions and nitrogen-containing ions in the low mass range below m/z 200. TOF-SIMS spectra of silk fabric dyed with curcumin, indigo, brazilin, shikonin and quercetin are shown in Figure 6. In the silk fabric dyed with curcumin, indigo, brazilin, shikonin and quercetin, molecular ions (M+H or M^+) at m/z 369, 263, 286, 289 and 303 were observed with strong intensity. The specific fragment ions from the dye molecules were also identified in the complicated background peaks from the silk textile.

Silk textiles were prepared with plant dye extracts and mordants or fixing agents in the dyeing process, because many of natural dyes did not have a strong chemical affinity for the textile fibers. As shown in Figure 7, TOF-SIMS spectra were obtained for the silk textiles dyed with sappanwood extracts and different kinds of mordants. In the low mass range of m/z 0–100, intense element ions from metal mordants were easily observed. Other metal ions in TOF-SIMS spectra were observed at the background level that might come from the contamination. The presence of aluminum, iron, tin and copper ion was due to the addition of mordants during the dyeing process of the silk fabric. High mass resolution TOF-SIMS spectra provided the characteristic ions from most mordant elements and dye molecules, separated with many hydrocarbon peaks. Therefore, TOF-SIMS is a useful technique to investigate natural dyes.

Silk fabric were dyed with various dye extracts of plants. Representative TOF-SIMS spectrum of silk textile dyed with indigo in the mass range of m/z 10–300 is shown in Figure 8. As can be seen in the figure, the specific fragment ions are present at m/z 235 and 247; as well, the molecular ion (M+H) of indigo can be observed at m/z

263. The specific zinc peaks from the mordants are also shown in this figure.

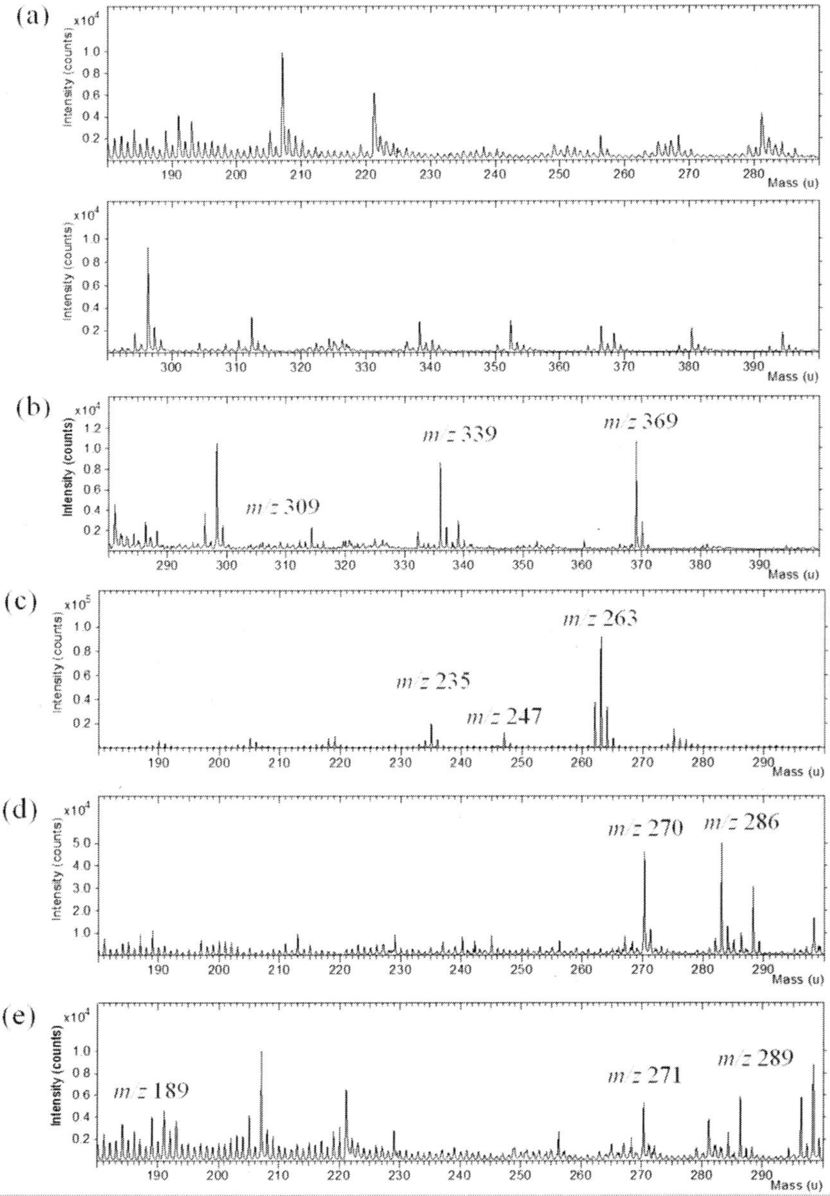

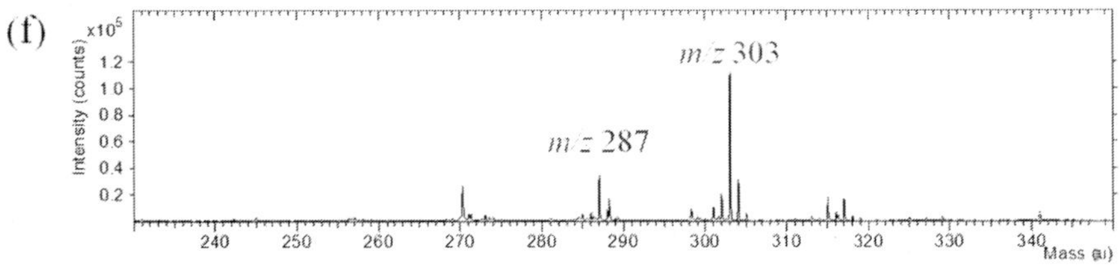

Figure 6. Positive ion TOF-SIMS spectra of silk fabric (**a**) undyed and dyed with commercial natural dyes: (**b**) curcumin; (**c**) indigo; (**d**) brazilin; (**e**) shikonin; (**f**) quercetin.

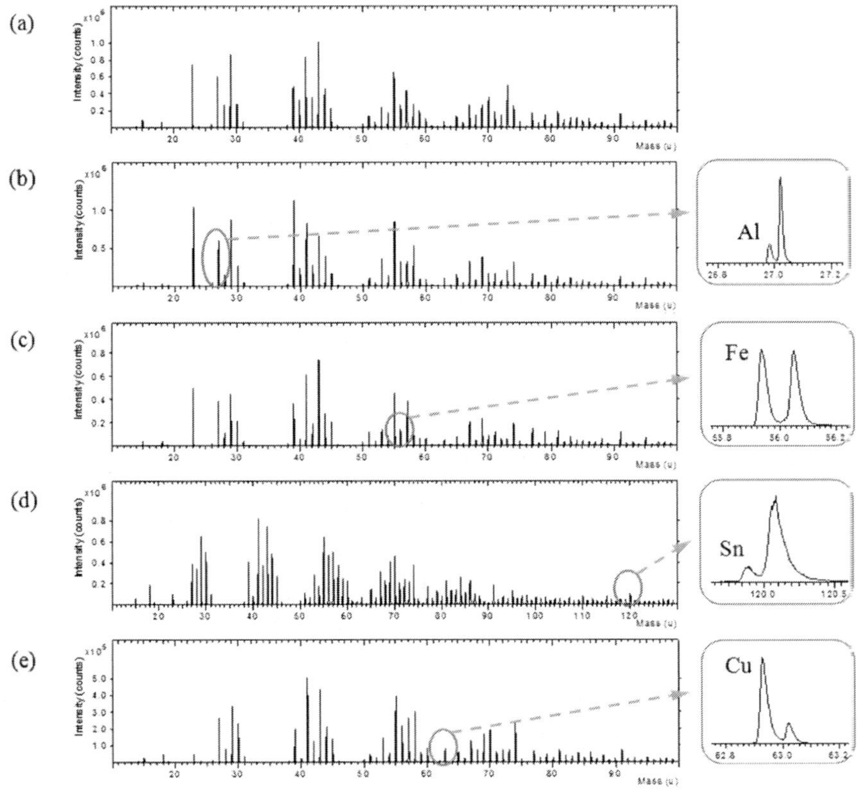

Figure 7. Positive ion TOF-SIMS spectra of sappanwood-dyed silk textiles (**a**) without mordant and with various mordants: (**b**) $KAl(SO_4)_2$; (**c**) $FeSO_4$; (**d**) $SnCl_4$; and (**e**) $CuSO_4$.

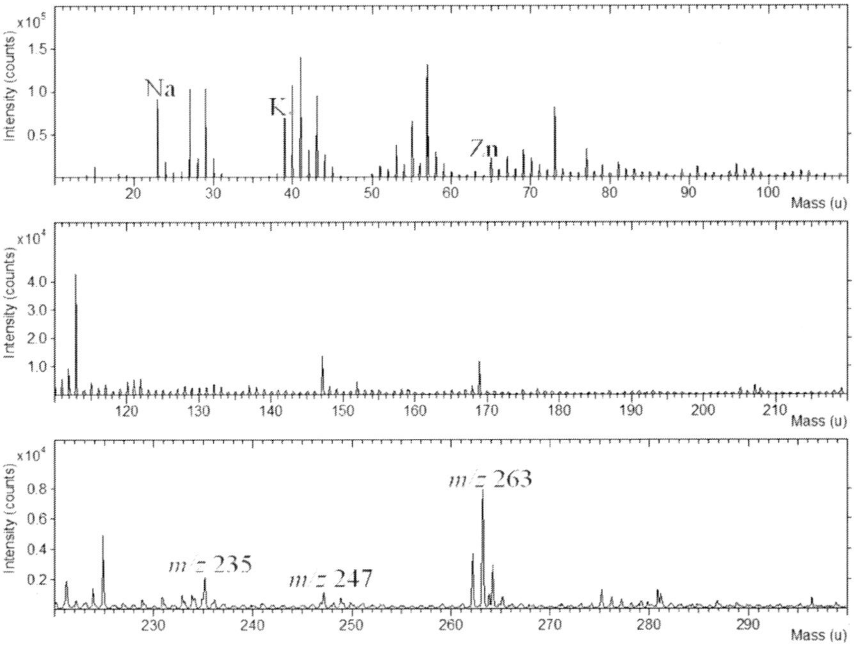

Figure 8. Positive ion TOF-SIMS spectrum of silk textile dyed with indigo plant extract and Zn powder.

The use of TOF-SIMS to characterize the chemical distribution of dye components on silk fabric was explored. TOF-SIMS imaging data was obtained for the silk textile dyed with sappanwood and $SnCl_4$. Figure 9 provides ion images obtained from the positive ion TOF-SIMS spectrum. No clear contrast between the dye distribution, and the silk textile surface is discernible. The TOF-SIMS images show the special distribution of Na^+, Sn^+, molecular ion ($C_{16}H_{14}O_5^+$, 286) and total ions. The distribution of the molecular ion exhibits a uniform distribution over the analyzed area, whereas the sodium and tin appear to be located in small aggregates. The TOF-SIMS image results indicate that most fractions of the silk fibers are covered by the plant extracted dye component during the traditional dyeing process according to the old recipes. However, metallic ion from the mordant agent was not distributed uniformly in the silk fibers. Sodium is not an element included in sappanwood component or the mordant. Sodium usually comes from the contamination of the sample and is easily ionized by the primary ion beam. The area contaminated by sodium was different in TOF-SIMS image map from those of dye component and mordent element.

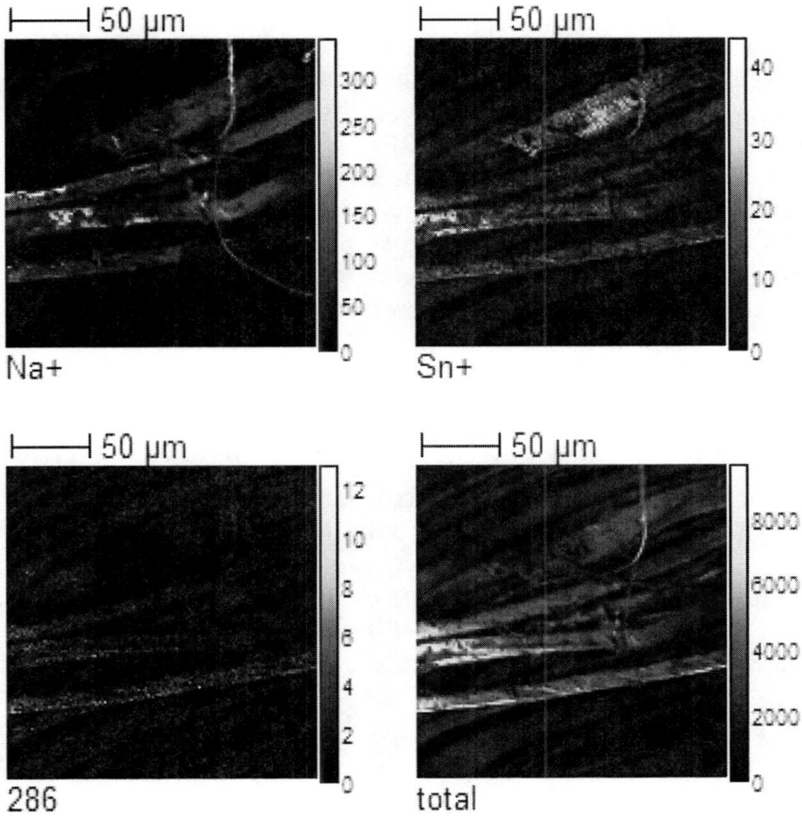

Figure 9. Positive TOF-SIMS images of sappanwood on silk textile: Na(m/z 23), Sn(m/z 120), brazilin(m/z 286) and total ion.

Therefore, plant extracted dyes on the silk textiles have been identified using the TOF-SIMS technique on a scale far below that which is possible when using the extractive technique. The TOF-SIMS technique provided information on both elements, including isotope discrimination and molecules, present on the surface of traditionally dyed silk fabric.

4. CONCLUSIONS

For a better understanding of Korean plant-extracted dyes and traditionally dyed textiles, the application of complementary noninvasive and micro-destructive surface techniques were established. Reference silk textiles were prepared with commercial natural dyes and were characterized by FTIR, XPS and TOF-SIMS. The analysis of silk fabric dyed with natural extracts from plants by traditional dyeing processes proved the usefulness of the reference data, which were compiled from the results of systematic testing on dyed silk fabric with commercial natural dyes. The FTIR technique did enable individual dye components of the dyeing materials in silk fabric to be distinguished, thus providing less possibility to discriminate between dyes belonging to the same chemical class and, therefore, not giving comprehensive information about the analyzed silk fabric. However, this technique can be well exploited as a preliminary scanning prior to sampling in order to minimize the sampling number.

Micro-destructive surface techniques, such as XPS and TOF-SIMS, led to the acquisition of detailed information on natural dyes, both in the reference silk fabric with commercial dyes and in silk textiles dyed with natural plant extracts by traditional process. The major advantages of XPS and TOF-SIMS to characterize the natural dyes and traditionally dyed textiles included small sample size requirements, minimal sample preparation, the lack of chemical or mechanical pretreatments and identification of the molecular structures of dye components. Quantitative results for various mordants in sappanwood-dyed silk fabric and for oxygen containing functional groups in traditionally dyed silk fabric were obtained by XPS technique. TOF-SIMS spectra of silk fabric dyed with various plant extracts were obtained in the mass range of m/z 0–500. Molecular ions and fragment ions from the plant-extracted dyes of traditionally dyed silk fabric were observed, as were elemental ions from the mordants. Establishment of surface analytical methods and systematic accumulation of reference data will provide a good understanding of traditional dyeing process in Korea.

ACKNOWLEDGEMENTS

This work was supported by the Korea Institute of Science & Technology (KIST) Institutional Program.

REFERENCES

1 Degano, I.; Robechini, E.; Modugno, F.; Colombini, M.P. Analytical methods for the characterization of organic dyes in artworks and in historical textiles. *Appl. Spectrosc. Rev.* **2009**, *44*, 363–410.

2 Little, S. Dye analysis helped to determine the treatment of two historically important textiles. *Dyes Hist. Archaeol.***2005**, *20*, 80–88.

3 Ferreira, E.S.B.; Quye, A.; McNab, H.; Hulme, A. The natural constituents of historical textile dyes. *Chem. Soc. Rev.***2004**, *33*, 329–336.

4 Rosenberg, E. Characterization of historical organic dyestuffs by liquid chromatography-mass spectrometry. *Anal. Bioanal. Chem.* **2008**, *391*, 33–57.

5 Nowik, W.; Desrosiers, S.; Surowiec, I.; Trojanowicz, M. The analysis of dyestuffs from first-to second-century textile artifacts found in the Martres-de-Veyre (France) excavations. *Archaeometry* **2005**, *47*, 835–848.

6 Ahn, C.; Obendorf, S.K. Dyes on archaeological textiles: Analyzing alizarin and its degradation products. *Text. Rev. J.***2004**, *74*, 949–954.

7 Clementi, C.; Nowik, W.; Romani, A.; Cibin, F.; Favaro, G. A Spectrophotometric and chromatographic approach to the study of ageing of madder (*Rubia tinctorum L.*) dyestuff on wool. *Anal. Chim. Acta* **2007**, *596*, 46–54.

8 Koren, Z.C. Historico-Chemical Analysis of Plant Dyestuffs Used in Textiles from Ancient Israel. In *Archaeological Chemistry*; Orna,

M.V., Ed.; ACS Symposium Series 625; American Chemical Society: Washington, DC, USA, 1996; pp. 269–310.

9 Petroviciu, I.; Wouters, J. Analysis of natural dyes from romanian 19th and 20th century ethnographical textiles by DAD-HPLC. *Dyes Hist. Archaeol.* **2002**, *18*, 57–62.

10 Zhang, X.; Good, I.; Laursen, R. Characterization of dyestuffs in ancient textiles from Xinjiang. *J. Archaeol. Sci.* **2008**,*35*, 1095–1103.

11 Wouters, J. A new method for the analysis of blue and purple dyes in textiles. *Dyes Hist. Archaeol.* **1992**, *10*, 17–21.

12 Surowiec, I.; Quye, A.; Trojanowicz, M. Liquid chromatography determination of natural dyes in extracts from historical Scottish textiles excavated from peat bogs. *J. Chromatogr. A* **2006**, *1112*, 209–217.

13 Sanyova, J. Mild extraction of dyes by hydrofluoric acid in routine analysis of historical paint microsamples.*Microchim. Acta* **2008**, *162*, 361–370.

14 Zhang, X.; Laursen, R.A. Development of mild extraction methods for the analysis of natural dues in textiles of historical Interest using LC-diode array detector-MS. *Anal. Chem.* **2005**, *77*, 2022–2025.

15 Wyplosz, N.; Heeren, R.M.A.; van Rooij, G.; Boon, J.J. Analysis of natural organic pigments by laser desorption mass spectrometry (LDMS): Preliminary study to spatially resolved mass spectrometry. *Dyes Hist. Archaeol.* **1998**, *16–17*, 187–198.

16 Van Elslande, E.; Guerineau, V.; Thirioux, V.; Richard, G.; Richardin, P.; Laprevote, O.; Hussler, G.; Walter, P. Analysis of ancient greco-roman cosmetic materials using laser desorption ionization and electrospray ionization mass spectrometry. *Anal. Bioanal. Chem.* **2008**, *390*, 1873–1879.

17 Lee, Y.; Lee, J.; Kim, Y.; Choi, S.; Ham, S.W.; Kim, K.J. Investigation of natural dyes and ancient textiles from korea using TOF-SIMS. *Appl. Surf. Sci.* **2008**, *255*, 1033–1036.

18 Andreotti, A.; Bonaduce, I.; Colombini, M.P.; Ribechini, E. Characterization of natural indigo and shell-fish purple by mass spectrometric techniques. *Rapid Commun. Mass Spectrom.* **2004**, *18*, 1213–1220.

19 Van Vaeck, L.; Adriaens, A.; Gijbels, R. Static secondary ion mass spectrometry: (S-SIMS) Part 1. methodology and structural interpretation. *Mass Spectrom. Rev.* **1999**, *18*, 1–47.

20 Lambert, J.B.; McLaughlin, C.D.; Shawl, C.E.; Xue, L. X-ray photoelectron spectroscopy and archaeology. *Anal. Chem.***1999**, *71*, 614A–620A.

21 Gillard, R.D.; Hardman, S.M.; Thomas, R.G.; Watkinson, D.E. The detection of dyes by FTIR microscopy. *Stud. Converv.* **1994**, *39*, 187–192.

22 Choo, C.K.K.; Lee, Y.E. Analysis of dyeings produced by traditional korean methods using colorants from plant extracts. *Color. Technol.* **2002**, *118*, 35–45.

CHAPTER 4

Materials and manufacturing environmental sustainability evaluation of apparel product: knitted T-shirt case study

Md Mashiur Rahman Khan and Md Mazedul Islam

Department of Apparel Engineering, Bangladesh University of Textiles, Tejgaon Dhaka-1208, Bangladesh

ABSTRACT

In this paper, evaluation of base materials and manufacturing environmental sustainability of selected branded T-shirt products made by Bangladesh is presented. The study is based on performing various eco-friendliness tests for the product base materials and evaluation though a standard tool named Higg Index. For this study, selected apparel branded T-shirt products from renowned brand, namely B, C, D, E, and F, and a researcher-developed BUTex-Innovation brand A are taken into consideration. Better environmental sustainability results, i.e., higher score, is obtained by A, B, C, and D branded T-shirts, but local branded product of E and F resulted in poor environmental sustainability with lower scores in terms of product eco-friendliness tests and High Index assessment tool. Besides, many weaknesses and opportunities for improvement of environmental sustainability in materials and manufacturing stages are identified and suggested to lead the textiles and clothing sectors towards sustainable growth.

KEYWORDS

Clothing products, Base materials, Knitted T-shirt, Higg index, Environmental sustainability

1. BACKGROUND

Sustainability is the major concern in the age of the modern world. For the textile and apparel sector, it has been a burning issue for many related concerned bodies. Challenges faced by the apparel manufacturers and retailers, brands with many incidents such as the Rana Plaza building collapse, the Tazreen garments fire tragedy in Bangladesh, and the crisis in Cambodia, led the clothing industry giants, nonprofit organizations, fashion designers, manufacturers, and consumers to be well aware towards sustainable textiles and clothing business practices. Over the past few years, increasing awareness of the environmental and social concerns surrounding the fashion industries and consumers has led to a rise in the implementation of sustainability initiatives. There has been a growing concern over apparel brands in improving their environmental impact and the social responsibility throughout their supply chains (Allwood et al. 2006). Most of the textiles and clothing product used by the consumer has a significant impact on the environment. For example, according to a study, one 0.5-lb cotton T-shirt requires 700 gallons of water during processing, uses 0.2 lb of fertilizers and 0.01-lb pesticides, and emits 6 lb of CO_2, 1.2 lb of fossil fuels, and 0.11 lb of other gases. One 0.5-lb pound T-shirt does 18.3 pounds of CO_2 emission on average from washing and dyeing 50 times in US (Fletcher and Grose 2012; USagain, 2015). Many consumers today do not know the extent to which these products impact the environment—low or high. They must account for the environmental impacts of the materials she uses, the resulting waste of the forms she chooses, how products are produced and packaged, where they will be made and then sold, causing the energy use for transportation, how consumers may use and dispose of the product (Adams and Frost 2008). A product is thought to be eco-sustainable if it is manufactured, used, and disposed of in a manner that would not harm the environment when compared to a product that was manufactured and used without any environmental concern. However, consumers have become more conscious of these impacts and are spelling out their preferences for eco-friendliness, thereby forcing the manufacturer to adopt clean technologies all along the supply chain to produce environmentally friendly products (Challa 2014). Environmental sustainability refers to the ability of something to continue without upsetting the earth's ecological balance. Sustainable apparel products can be defined as a part of the growing design philosophy and trend of sustainability, the goal of which is to create a system which can be supported indefinitely in terms of environmentalism and social responsibility (Adams and Frost 2008). Environmental sustainability in business refers to longevity but in terms of which natural resources the production process might draw upon, how resources are used and replenished, the overall impact of the final production on the environment, and where the product ends up following its disposal. The textile and apparel products impact the environment at every point along the product's lifecycle (Fletcher 2008). However, rapid growth in the textiles and clothing industry sector is playing a vital role in the economy of Bangladesh. Despite the growing necessity of environmental practices in the apparel industry, little research has explored methods and standard tools to guide apparel designers, manufacturers, merchandisers, and consumers in adapting environmental issues in their works. Many designers do not recognize how their designs impact the environment. The role of the designer must expand dramatically in light of current environmental concerns with the contamination and destruction of the ecosystem (DEFRA 2010). Today's designer, manufacturer, merchandiser, and consumer must be well informed on a vast array of topics, from methods of production to governmental regulations to life cycle analysis. They must account for the environmental impacts of the materials' uses, the resulting waste of the forms they choose, how products are produced and packaged, where they will be made and then sold, the causing energy use for transportation, and how consumers may use and dispose of the product (Gowork 2011). All products that are manufactured cause environmental degradation, either during manufacture, use, or disposal. This can be evaluated by looking at the different phases of the product's life cycle and taking action at the phases where it will be most effective to reduce the environmental impact. Among many techniques to study the eco-impact of products, the life cycle assessment is most widely used one. LCA examines products from initial stage to the final stage, covers the entire life cycle, and also examines eco-impact during the whole lifetime (Muthu et al. 2012). However, the life cycle of a product is long and complicated, covering many areas with many people involved in each phase. A remedial measure or policy may not be possible to address this aspect, but a variety of voluntary and mandatory tools will help to achieve this objective. These include eco-friendly tests for related clothing products, economic instruments, bans on certain substances, environmental labeling, voluntary agreements, and product design guidelines (IPP 2014). Hence, there is a growing need to evaluate the environmental sustainability of apparel products by standard methods, tools, and techniques to minimize the environmental impact. The Higg Index by the Sustainable Apparel Coalition is such a standard tool by which the environmental sustainability of apparel products could be evaluated. This study attempts to identify and analyze a number of these issues related to apparel product sustainability focusing on materials and manufacturing environmental sustainability by taking several local and foreign branded knitted T-shirts into consideration as a case study.

1.1 Literature review

The clothing industry is a popular industry among consumers, but it has great effects on many environmental, social, and governance concerns (Challa 2014). The textile industry prepares the base materials and the fashion, and the apparel industry converts these materials to suit the desires and needs of consumers; both industries are responsible for high utilization of energy, water, chemicals, and resources from cotton to petroleum (Niinimäki and Hassi 2011). The poor on-site conditions of the textile factories and working environment have caused many problems for the workers and operators, forming the basis for social reforms. In addition, the precarious supply chain upon which many manufacturers rely to develop apparel products can cause many problems for merchandisers and retailers for sustainable apparel product manufacturing.

1.2 Sustainable apparel product

Sustainability is defined as the design of human and industrial systems to ensure that humankind's use of natural resources and cycles do not lead to diminished quality of life due either to losses in future economic opportunities or to adverse impacts on social conditions, human health, and the environment (Lewis and Chem 2006). Environmental science defines sustainability as the quality of not being harmful to the environment or depleting natural resources and thereby supporting long-term ecological balance. The sustainable textile products can be defined as the textile products produced using raw materials, energy, various resources, and other ingredients which are derived from renewable resources that cannot be exhausted and consequently do not affect the next generation (Muthu et al. 2010). To achieve sustainable development, designers need to be aware of environmental impacts and incorporate environmental awareness into the design (Adams and Frost 2008). Product sustainability is the easiest aspect to alter for an apparel brand, as this is where a company has the most and direct control through design and product development. Transforming product sustainability may be achieved via various aspects such as fiber/textile selection, processing methods, use behaviors, and reuse/recycle strategies. Material selection is often the first step that designers and product developers will take in reducing the environmental impact of the garment. Environmentally preferred fibers/textiles can significantly reduce the environmental impact and increase the resourcefulness of an apparel product throughout the garment life cycle without changes to design practice or product development processes (Allwood et al. 2006).

1.3 Environmental impact of textiles and clothing products

The apparel industry has many negative environmental and social impacts that are complex and occur at different stages of the apparel life cycle. The development of fast fashion has amplified the impacts due to the increased volume of apparel produced and sold at low prices (Laursen et al. 2007). The apparel has a long and complicated production chain consisting of many phases including resource production and extraction, fiber and yarn manufacturing, textile manufacturing, apparel assembly, packaging, transportation and distribution, consumer use, recycling, and ultimate disposal (Islam and Khan 2014). The environmental impacts of apparel are varied across the phases, difficult to assess for individual garments, and are subject to the type of raw material used, dyeing and laundering. The major environmental impacts associated with the production and use of apparel throughout its life cycle include wastewater emissions from dyeing, finishing, and washing processes, increase in pollution, solid waste production, and significant depletion of resources from consumption of water, fossil fuels, and raw materials (Islam et al. 2014). Energy is used for laundering, transportation, operations of machines for various processes, the production of primary materials, especially man-made fibers such as polyester (a petroleum-based product), and yarn manufacturing of natural fibers such as cotton (Laitala and Boks 2012). Conventional cotton production has high water consumption and employs the use of toxic chemicals that may harm human health and the environment (Allwood et al. 2006). Chemicals are also released in wastewater from processes such as pre-treatments, dyeing, finishing, and laundry. These chemicals are disruptive to both the environment and aquatic-based life (Ruskino 2007). Solid waste is produced during natural fiber yarn, textile, and apparel manufacturing and disposal of apparel products at the end of their life. There are significant issues with clothing waste as the majority of clothing and textile wastes ends up in landfills as opposed to being recycled or reused (Goworek 2011). However, environmental impacts cannot be addressed once the product is introduced into the market and more attention should be focused on eco-environmentally friendly product design. Improvements in product design can be made by improving the flow of life cycle information and eco-design guidelines, integrating environmental considerations into the manufacturing processes, and involving relevant stakeholders to review the approach (Steinberger et al. 2009). Education on the need for eco-friendly products and the use of case studies and examples would help to strengthen the design and manufacture of environmentally sustainable products. European Commission-Health and Consumers, Scientific Committees Opinion (1999).

1.4 Impact of materials selected to eco-sustainability

Of all the life stages, the apparel industry has concentrated its environmentally responsible innovation in developing eco-friendly materials, for some understandable reasons. Material choices incorporate the environmental impacts of the production of those materials into the apparel product's life cycle and affect how the garments can be processed, cared for, and disposed (Lewis and Chem 2006). The environmental concerns for materials include energy consumption, use of toxic chemicals, ethics, and sustainability of resource consumption (Muthu et al.2009). In order to address these material environmental concerns, eco-friendly apparel designers must strive to minimize non-renewable resource and energy use, as well as waste, while utilizing materials that are nontoxic and ethically produced. These materials should also provide valuable characteristics to the consumer, require low-impact care, and be effectively disposed through reuse, recycling, energy reclamation, or composting (Goan 1996). The apparel designer's primary material is of course fabric that can be composed of various fiber types that have unique properties and production. The two principal fiber types are natural fibers from plant cellulose and animal proteins and manufactured fibers derived from natural and synthetic polymers. Fibers are either cultivated, manufactured, or a combination of both. Each fiber source and production method come with its own set of environmental concerns, yet all can be improved to various degrees of reduced environmental harm by either employing best practices in current systems, implementing new, more effective technologies, or replacing current materials and practices with eco-friendly alternatives.

1.5 Impact of manufacturing processes to environmental sustainability

Clothing production is a multi-step procedure that often includes spinning, weaving, knitting, pre-treating, dyeing and/or printing, finishing, and makeup (cut-sew-trim). During these stages, the apparel product is subjected to various chemical treatments, many of which have been highly toxic and non-degrading, but now are sometimes replaced by more nontoxic and biodegradable counterparts (Kadolph 2010). Some fiber processing standard exists that prohibit the use of hazardous chemicals, including the Organic Trade Association's (OTA), American Organic Standards for Fiber Processing (Murray and Coody 2003), and the Global Organic Textile Standard (GOTS). Each production stage demands energy and often water to apply treatments, which then require more energy to dry the product after treatment. And lastly, the working conditions at each stage can pose health and safety risks due to exposure to toxic chemicals and air and noise pollution. Eco-friendly apparel production uses resources and energy effectively, only uses nontoxic cleaners and applications. Water is used during many fabric production stages for cleaning and application of dyes and finishes, resulting in large amounts of problematic wastewater (Nese and Ismail 2007). This wastewater can contain washed away dirt and wax from natural fibers, cleaning chemicals, auxiliaries, etc. Most wastewater is dumped rather than treated due to the expense, causing disruptions in the local environment and contaminating the fresh water supply. Dyeing can be done at almost any stage, from the fiber to the constructed product, and may require whitening pre-treatment (Ruskino 2007). Dyes and printing inks require a pigment and a fixative, or mordant, both of which can utilize toxic chemicals and heavy metals. Azo group chemicals, which release carcinogenic arylamines, are widely used in synthetic dyes and pigments. Dyes are difficult to remove from wastewater and leave it colored, preventing wastewater from supporting plant life by inhibiting aquatic plants' ability to photosynthesize. Eco-friendly alternative color options include low-impact and natural dyes that use nontoxic mordants, water-based inks, and color-grown cotton and wool fibers, although the color-grown palates are often muted and limited (Challa 2014). Finishing occurs after the fabric is made and often implies the application of chemicals to the fabric to give it enhanced properties, either esthetic or functional. This process often requires water for the application and to wash out the excess finish. Many finishing processes have been redesigned to use less water or foam substitute, although they still need to be improved to reduce water use and increase wastewater processing. Toxic chemical finishes pose human health risks while their wastewater disrupts ecosystems, yet more finishes are being replaced with less toxic, biodegradable alternatives. Certain finishes can also require the garment to be "dry-clean only," which uses cleaning solvents with toxic chlorine chemicals. Conversely, finishes can make traditionally dry clean only fibers, like wool and silk, washable. Some, such washable finishes are toxic chemicals while other uses environmentally safe, natural enzymes (Chen and Burns 2006). Make-up, also known as cut-make-trim, involves cutting the garment pattern out of fabric, sewing it together, and then adding any trim or embellishments. By eliminating waste in the overall process, the product's manufacturing sustainability can be increased remarkably (Islam et al. 2014). Wastages from cutting can vary between 6 and 25 %, depending on the complexity of the pattern, although this cutting waste is often recycled into low-quality textile products (Laursen et al. 2007). The apparel industry still relies heavily on human labor as no machinery has been able to reproduce the agility and intricacy of a human producer.

1.6 Introduction to the Sustainable Apparel Coalition and Higg Index tool

The world's textile and apparel industry cover various areas that include the manufacturing, marketing, and retailing of textiles and garment products. This industry has been considered as an approach for industrialization, economic progress, and national development. According to the World Trade Organization, China has been leading the world with regard to export in the field of textiles and apparel, followed by the European Union and India (Wolf et al. 2011). Despite many recent recessions, strategic policies taken in this sector have saved the industry from various problems. Every industry should concentrate on keeping stock levels low, as well as on being flexible and in fine tune with the consumer's needs and wants, emphasizing lean management and strong supply chain networks. Ecological friendliness is the main motto and consumers have been very much aware of its impact, seeking out products that complied with ecological standards (Speer 2005).

However, the Sustainable Apparel Coalition (SAC), which helps industries to rate their products with a numerical sustainability, scores in order to provide data to customers with regard to the extent the manufacturer has contributed to the conservation of the environment. SAC (2013) The Higg Index is one such tool for the assessment of product's sustainability, which raises a manufacturer's consciousness of the design, choice of raw materials, manufacturing processes, and finishing, packaging, and distribution through the use and recycling of the product (Martin 2013; Reichard 2013). The SAC is a multi-stakeholder engagement, formed in 2011, by a group of global apparel and footwear companies and non-profit organizations representing nearly one third of the global market share for apparel and footwear. They seek to build a common approach for measuring and evaluating apparel and footwear product sustainability performance (Ulibarri 2011). It aims to develop common measurements and a common environmental understanding of a product impact across the industry by building on the Outdoor Industry Association (OIA)'s, Eco Index™, and Nike's Environmental Design tool (Nike Inc. 2012c). The Eco Index™ is a standardized tool for measuring the environmental impacts of outdoor products such as clothing and tents and evaluates the impacts in six key areas of a product's life cycle, namely materials, packaging, product manufacturing and assembly, transport and distribution, use of services, and end of life (Nike Inc. 2012c). Measuring environmental performances of apparel products will spotlight priorities for action and opportunities for technological innovation. SAC (2013c) The Higg Index is an "indicator based tool for the apparel that enables companies to evaluate material types, products, facilities and processes based on a range of environmental product design choices" (The Higg Index 2012). The Higg Index is a set of assessment tools that are used to evaluate the environmental impact of apparel and footwear products. The index was initially released on 26 June 2012 as the Higg Index 1.0, and it has been used by many organizations—both SAC members and nonmembers. The Higg Index 1.0 used a Microsoft Excel interface and worked on qualitative indicators for assessment. The sustainability topics were related to the environment, and the product category was apparel. Approximately 44 materials were included in the material assessment data, and the material sustainability index used basic indicator questions on the environment. The Higg Index version 2.0 is the updated version of 1.0 and was released on 11 December 2013. It is based on tools such as the Eco Index, Nike's Apparel Environmental Design Tool, Global Social Compliance Program reference tools, and social/labor practice tools. After a pilot testing period and use of the second version by many organizations for over 14 months, the Higg Index 2.0 was introduced SAC (2013b). This tool helps to standardize the methodology for measurement and evaluation of the environmental performance of the apparel products along the supply chain in three levels—namely the brand, product, and facility levels. The scope of the Higg Index 2.0 is to assess the environmental and social/labor performance of apparel and footwear products. It is based on life cycle analysis spanning the entire lifecycle of apparel products, encompassing raw materials, manufacturing, packaging, transportation, use, and end of life. The Higg Index 2.0 is a tool that educates small and large companies to recognize challenges and sustain improvement. The self-assessment tool helps scientific learning by means of identifying the vital aspects of environmental sustainability and opportunities that will provide improvement. This index is the basis for future operations and efforts to ensure sustainability and also the starting point for the commitment, learning, and collaboration among stakeholders. The scope of the desired outcomes of the SAC Higg Index tool includes improvement to reduce water use and improve quality, reduce energy and emissions, minimize waste, reduce chemicals and toxicity, and increase transparency of social and labor issues (Ruskino 2007). The tool evaluates material type, products, facilities, and processes. By utilizing practice based, qualitative binary yes/no questions, assessment can be made as to the sustainability performance of the product and drive behavior for improvement. The Higg Index RDM-Beta is a prototype which aims at educating and providing quick directional guidance to apparel designers during the product creation process about the potential environmental impacts of their design solutions (SAC 2013a). It is to engage designers in the Higg Index's product life cycle thinking and how we directionally assess materials sustainability through the MSI. Through self-assessment, organizations can better understand the environmental impacts that occur throughout the life cycle and the effects of design choices such as material type SAC (2012i). However, there are many apparel brands belonging to the Higg Index such as: H&M, Gap Inc., Nike, Adidas, Puma, Patagonia, Mountain Equipment Co-op (MEC), Levi's, Hanesbrand, Marks and Spencer, Esprit, Columbia, Timberland, Loomstate, S. Oliver, and many such bold face names. These brands are widely using this Higg Index tool to measure environmental performance of their apparel products towards better sustainability of their products (Reuben 2013).

2. METHODS

To carry out this study, selected knitted men's T-shirt products of 100 % cotton single jersey fabric was taken into consideration as a case study. Both foreign and local branded knitted men's "M" sized short sleeve T-shirt products from some renowned brand were selected on a random basis to measure their environmental performance. The brand B-, C-, and D-named T-shirts were collected from an apparel industry named Viyellatex Group Ltd. T-shirts from brand E and F were collected from Beximco fashion wear and Aarong. BUTex-Innovation T-shirt was developed by the researcher as eco-friendly product which is indicated as brand A in this study. However, environmental sustainability of mentioned branded T-shirt was evaluated based on mainly two aspects only. One is to perform selected eco-friendly tests and another is an evaluation of the products by SAC Higg Index tools. Although, some other aspects are related to a fully environmentally sustainable apparel product.

2.1 Product environmental aspect test procedures

To evaluate environmental sustainability of base materials, an initiative was taken to carry out the eco-friendly test of product base body fabric that included formaldehyde, pH value, analysis of amines in azo dyestuff content, and alkylphenol ethoxylate (APEO) content test. These entire eco-friendly tests were performed as per standard test procedure. The standard test procedure followed for various test were the following: for formaldehydes - International Standard ISO 14184–1: 2011 (for textile and non-woven), for pH-ISO 3071:2006 extraction with potassium chloride (for textiles), for analysis of amines in azo dyestuffs - European Standard EN 14362–3: 2012, (For textile) and for alkylphenol ethoxylate (APEO) content - methanol extraction and analysis of by liquid chromatograph mass spectrometer (LC-MS). If the test results are out of range expected as per standard, then that products is termed as harmful to environmental sustainability in that aspect.

2.2 T-shirt materials and manufacturing environmental aspect evaluation by SAC Higg Index tool

All the six T-shirt products were evaluated by the SAC Higg Index tool that provides a value that represents how environmentally responsible an apparel item is in terms of sustainability. For this study, the tool Higg Index was used to obtain product scores for the abovementioned T-shirts. The final score was obtained by answering questions relating to the product materials and manufacturing environmental sustainability issues. The data used during this process include information readily available to the manufacturers, suppliers, merchandisers including hang tags, care labels, and product technical package descriptions, and throughout the LCA phase of the T-shirt. The product technical package detail information was the important source for various Higg Index-required information towards eco-sustainability evaluation. Higg Index requires that all materials present be entered into a material input table along with eco-manufacturing issues. This includes a description of all materials used in addition to information obtained from follow-up questions. By utilizing practice-based, qualitative binary yes/no questions, assessments were made as the sustainability performance of the product and drive behavior for improvement. Higg Index is a Microsoft Office Excel and web-based tool that has predetermined standard values based on a global survey for specific materials that are used in determining a score. Various links between inputs and outputs considered were based on Higg Index tools as the tool has predetermined standard values based on global survey and mathematical model for specific materials and manufacturing process issues. The higher the Higg Index score, the better the environmental performance of an apparel products.

3. RESULTS AND DISCUSSION

In this study, the all five selected foreign and local branded knitted T-shirts along with newly developed one were assessed using the Higg Index 1.0 tools to obtain a value that represents a product's environmental sustainability especially in material and manufacturing issues. A higher score implies a more sustainable product, and a lower score implies less sustainability. Upon completion of the excel-based Higg Index tool, the brand A, B, C, D, E, and F T-shirts received different scores on a scale of 1–100. The superior score of the BUTex-Innovation (brand A) and the foreign branded products, especially the brand B-shirt, was due to the organic material content, eco-friendly raw materials and process used product end of life options and confirmed traceability. On the other hand, the local branded T-shirt gained comparatively lower score in most of the areas due to not using eco-friendly raw materials, manufacturing processes, high-impact finishes, improper or insufficient care and repair instructions, and traceability issues. This also assisted to find out various gaps regarding apparel product materials and manufacturing issues and to take steps to improvement of superior clothing product in terms of sustainability.

3.1 Results for environmental evaluation test of T-shirt product

Environmental issues are playing an increasingly important role in consumer products. Textiles and clothing are also subjected to stringent examination, particularly with regard to the chemicals used in their manufacture as they are often in direct contact with the skin. The consumer is confronted with scare stories of health risks and pollution allegedly originating from modern textiles treated with "poisonous, allergenic, or carcinogenic dyes and finishing auxiliaries". Such gross misrepresentation is completely untrue for textile production in countries, where extensive environmental and chemical pollution laws exist (GB 18401 2010). Nevertheless, it cannot be ruled out that some textiles of uncertain origin traded on international markets may have been processed inadequately and may contain residues of untested, possibly hazardous, auxiliaries and dyes. In light of this, the responsible partners in the textile industry believe that it is important to reassure consumers about textiles in human ecological terms, i.e., to reassure the public that properly processed textile products present no health risks (GB 18401, 2010). However, various eco-friendly tests performed for the above selected T-shirt product item showed significant difference.

3.2 Formaldehyde test

Formaldehyde has allergenic and sensitizing effects, even at low concentrations. Levels of 300-mg/kg formaldehyde on a garment can trigger allergic effects in sensitized individuals. Formaldehyde has an irritant effect on the respiratory tract and mucous membranes in concentrations above 0.5 mg/m^3 air. Test method followed for formaldehyde is the International Standard ISO 14184–1: 2011 (for textile and non-woven). The T-shirts of A, B, C, and D brands showed better results in formaldehyde detection with value 11.42, 20.47, 17.82, 32.45 ppm, respectively. But a local branded T-shirt of E and F brand showed poor results with higher value, i.e., 46.23 and 62.35 ppm, respectively. Results are shown in Table 1.

Table 1 Results of product eco-friendliness test for various selected brand

Branded T-shirt	Results of formaldehyde test (in ppm). (Detection limit: 16 ppm) (Maximum limit: 75 ppm)	pH value test (Maximum limit: 4.0–7.5 ppm)	Results of analysis of amines in azo dyestuffs (in ppm) (Detection limit: 10 ppm)	Results of alkylphenol ethoxylate (APEO) content test in ppm (Detection limit: 30 ppm) (Maximum limit: 100 ppm)
Brand A	11.42	6.4	Not detected	Not detected
Brand B	20.47	5.7	Not detected	Not detected
Brand C	17.82	6.2	Not detected	Not detected
Brand D	32.45	5.5	13.5	Not detected
Brand E	46.23	7.4	31.7	47.6
Brand F	62.35	7.5	43.2	64.3

Various test methods—formaldehyde test method: International Standard ISO 14184-1: 2011 (for textile and non-woven); pH value test method: ISO 3071:2006 extraction with potassium chloride (for textiles); analysis of amines in azo dyestuff test method: European Standard EN 14362-3: 2012. (for textile); alkylphenol ethoxylate (APEO) content test method: methanol extraction and analysis of by liquid chromatograph mass spectrometer (LC-MS) (for textile)

3.3 pH value test

The pH value of an aqueous extract of the textile article is determined according to ISO 3071. The required pH range corresponds approximately to the natural pH value of human skin (approximately 5.5). This avoids skin irritation due to the release of acidic or alkaline substances from damp textiles. The pH maximum limit for safe textile materials is 4.0–7.5. However, the tested pH value results for various branded T-shirts were 6.4, 5.7, 6.2, 5.5, 7.4, and 7.5 for brand A, B, C, D, E, and F T-shirts, respectively (Table 1). Also, noticed that the pH value for local brand E and F T-shirts was above 7.0.

3.4 Analysis of amines in azo dyestuffs

Azo dyes are compounds characterized by their vivid colors and provide excellent coloring properties. They are important and widely used as coloring agents in the textile and leather industries. The risk in the use of azo dyes arises mainly from the breakdown products that can be created in vivo by reductive cleavage of the azo group into aromatic amines. Due to the toxicity, carcinogenicity, and potential mutagenicity of thus formed aromatic amines, the use of certain azo dyes as textile colorants and the exposure of consumers using the textile colored with azo compounds cause a serious health concern. The two main routes of consumer exposure are the skin absorption of the azo compounds from the dyed clothes worn and potential oral ingestion, mainly referring to the sacking of textiles by babies and young children. In this study, test method followed by analysis of amines in azo dyestuffs is the European Standard EN 14362–3: 2012 (for textile). From Table 1, it is seen that the A, B, and C branded T-shirts showed better results i.e., no detection of such substances, but in D, E, and F branded T-shirts, these substances were detected with values 13.5, 31.7, and 43.2 ppm, respectively. Use of eco-friendly raw materials and third party certified dyes and chemicals may add to better results for A, B, and C branded T-shirts.

3.5 Product test for alkylphenol ethoxylate (APEO) content

Analyses for APEO have been a fast growing issue of textile material testing. Surprisingly to the producers, APEOs have been detected in rather high amounts in textiles. The most important APEOs or alkylphenol ethoxylates for the textile industry are NPEOs (nonylphenol ethoxylates) and OPEOs (octylphenol ethoxylates) due to their detergent properties. About 90 % of the produced APEOs are in fact NPEOs. The compounds are used in detergents, cleaning agents, or chemicals used for textile or leather production. Toxicological characteristics are their hormone-disruptive properties and the fact that they are toxic to aquatic organisms (Ljungberg 2007). Furthermore, APEOs are very persistent and difficult to be degradable in nature. These are all problem areas for wastewater treatment and discharging of wastewater into surface water. In this study, standard test method followed for textile is methanol extraction and analysis of by liquid chromatograph mass spectrometer (LC-MS). No detection of APEO was found for A, B, C, and D branded T-shirts due eco-friendly dyes and chemicals used in manufacturing processes. Alternatively, a value of 47.6 and 67.3 ppm was detected for E and F branded T-shirt test which might be due to environmentally harmful dyes and chemicals and raw material selection in manufacturing stages. Table 1 presents the result details for APEO test.

3.6 Test results for various branded T-shirts

3.6.1 Higg Index evaluation results for materials

The materials selected for the knitted T-shirt for both foreign and local branded products were entered into the material input table in the Higg Index tool. The values obtained by various products are shown in Table 2. The developed men's T-shirt A and foreign brand B, C, and D obtained higher score. This was due to the eco-friendly raw materials used, i.e., brand A and B T-shirt included 100 % organic cotton fiber. During processing of fabric, BUTex-Innovation T-shirt, B-, and other foreign branded T-shirt used eco-friendly, third party verified and environmentally low-impact dyes and chemicals. But in the case of local branded E and F T-shirt, the Higg Index score was lower due to environmentally hazardous conventional raw materials, processing, poor quality dyes used, no third party verification and such types, and many other issues. For better environmental sustainability, these issues must be addressed for local branded apparel products. Obtained score out of 100 points by A, B, C, D, E, and F branded T-shirts are 63.3, 58.8, 51.6, 46.6, 39.6, and 30.6, respectively. Major variation issues in scoring of branded T-shirt by Higg Index are shown in Tables 3 and 4.

The material sustainability index (MSI) also examines chemistry, energy, and GHG intensity, water- and land-use intensity, and physical waste across the entire cradle-to-gate life cycle of a material. Land-use intensity focuses solely on the origin of a raw material in phase 1 and does not consider land use elsewhere in the material life cycle. The MSI compiles and scores data for 100 % cotton knit (For A, B, C, D, E, and F branded products) fabric based on standard global survey data compiled and established by 13 individual indicators within the various impact areas (MSI 2013).

Table 2 Total materials sustainability score for A and B branded T-shirts

Materials eco-sustainability scoring issues	Resultant answer (For A and B brand)	Obtained materials Score points (A)	Obtained materials Score points (B)
Materials name	Fabric (100 % organic cotton knit fabric	28.2 (out of 50)	28.2 (out of 50)
Certified recycled content %	0 %	0.0	0.0
Virgin, certified organic content %	100 %	10.0	10.0
Other source certification or verified chain of custody with name	Yes, GOTS, BCI certified	7.0	7.0
Coating/laminate finish? (If yes, specify type)	No	4.0	4.0
Reduced water use in dyeing/coloring? (If yes, specify reduced water process/technology)	Yes, used process or technology that reduces water consumption (and reduction is documented), eco-friendly dyes and chemicals used, enzymatic scoring used	5.0	5.0
Chemical impact reduction ("green chemistry") considered?	Yes	4.5	4.5
Green chemistry third party verification?	"Yes" OEKO-Tex Standard-100 (class-II) verified) but "No" for BUTex-Innovation T-shirt	0.0	4.5
		58.8 (out of 100)	Total = 3.3 (out of 100)

Table 3 Total materials sustainability score for C and D branded T-shirts

Materials eco-sustainability scoring issues	Resultant answer		Obtained materials Score points (C)	Obtained materials Score points (D)
	C branded T-shirt	D branded T-shirt		
Materials name	Fabric (100 % cotton knit fabric	Fabric (100 % cotton knit fabric	26.6 (out of 50)	26.6 (out of 50)
Recycled content %	0 %	0 %	0.0	0.0
Virgin, certified organic content %	0 %	0 %	0.0	0.0
Other source certification or verified chain of custody with	Yes, GOTS, BCI certified	Yes, BCI certified	7.0	7.0
Coating/laminate finish? (If yes, specify type)	No	No	4.0	4.0
Reduced water use in dyeing/coloring? (If yes, specify reduced water process/technology)	Yes, used process or technology that reduces water consumption (eco-friendly dyes and chemicals, enzymatic scoring used)	No, used traditional dyeing/ coloring process or technology	5.0	0.0
Chemical impact reduction ("green chemistry")?	Yes	Yes	4.5	4.5
Green chemistry third party verification?	Yes, OEKO-Tex Standard-100 (class-II) verified)	Yes, OEKO-Tex Standard-100 (class-II) verified)	4.5	4.5
			Total = 51.6 (out of 100)	Total = 46.6 (out of 100)

Table 4 Total materials sustainability score for E and F branded T-shirts

Materials eco-sustainability scoring issues	Resultant answer		Obtained materials Score points (E)	Obtained materials Score points (F)
	E branded T-shirt	F branded T-shirt		
Materials name	Fabric (100 % cotton knit fabric	Fabric (100 % cotton knit fabric	26.6 (out of 50)	26.6 (out of 50)
Recycled content %	0 %	0 %	0.0	0.0
Virgin, certified organic content %	0 %	0 %	0.0	0.0
Other source certification or verified chain custody and name	No	No	0.0	0.0
Coating/laminate finish? (specify type)	No	No	4.0	4.0
Reduced water use in dyeing/coloring? (if yes, specify reduced water process/technology)	No, used traditional dyeing/coloring process or technology	No, used traditional dyeing/coloring process or technology	0.0	0.0
Chemical impact reduction ("green chemistry")?	Yes	No	4.5	0.0
Green chemistry third party verification?	Yes, OEKO-Tex Standard-100 (class-II) verified)	No	4.5	0.0
			Total = 39.6 (out of 100)	Total = 30.6 (out of 100)

3.6.2 Higg Index evaluation results for manufacturing stages

The manufacturing stage is obviously an important area for better product sustainability. Some major issues have huge impact on product environmental sustainability as per Higg Index tool. The developed men's T-shirt, B-, and other foreign branded T-shirt used low-impact garment finish, i.e., process combination enzyme wash or basic rinse that scored higher. But local branded F and E T-shirts used high-impact garment finish, namely traditional bleach and acid chemical with stone wash, and scored lower. Besides, marker efficiency has great impact on product sustainability. The higher the marker efficiency, the better the product sustainability; and lower marker efficiency indicates its direct impact on supply chain, including more wastage of material, fabrics, dyes, chemicals, process loss, land fill, etc. As developed BUTex-Innovation brand A, B, C, and D branded T-shirts had more than 91 % marker efficiency, they scored better. But E and F branded T-shirts had marker efficiency of 82 and 85 % and scored lower in the Higg Index tool. Similarly, a single colored, screen-printed logo with small area received a higher score for foreign branded T-shirts and these issues vice versa scored lower for local branded T-shirts. The total points for manufacturing stages by various branded T-shirts are 60, 72, 65, 63, 49, and 27 for A, B, C, D, E, and F respectively which is shown in Table 5.

3.7 Integrating issues for environmental sustainability improvement of apparel products

The researcher has identified many technical requirements and issues for apparel product sustainability improvement based on the Higg Index assessment tool.

Table 5 Higg Index evaluation on manufacturing eco-sustainability of various branded T-shirts

Manufacturing eco-sustainability of various branded T-shirts	Answers of questionnaires						Score Point Obtained (out of 100)					
	Brand A	Brand B	Brand C	Brand D	Brand E	Brand F	Brand A	Brand B	Brand C	Brand D	Brand E	Brand F
Marker/material efficiency %	91	93	95	93	85	82	41(out of 50)	43	45	43	30	24
Garment finish	Basic rinse	Enzyme wash	Basic softener	Basic softener	Basic softener	Acid chemical wash	10 (out of 30)	20	10	10	10	0
Applied graphics and logo	10 cm^2 or less	10 cm^2 or less	10 cm^2 or less	10 cm^2 or less	10–100 cm^2	100–200 cm^2	6.0 (out of 10)	6.0	6.0	6.0	6.0	2.0
Screen printing size and color	10–100 cm^2 and 3 color	10–100 cm^2 and 3 color	10 cm^2 or less and 2 color	10 cm^2 or less and 2 color	10–100 cm^2 and 3 color	100–200 cm^2 and 5 color	3.0 (out of 10)	3.0	4.0	4.0	3.0	1.0
Total							60	72	65	63	49	27

3.8 Sourcing fair production and practice transparency:

Sourcing fair production can be performed by selecting manufacturers that are known to operate in an eco-socially responsible manner. For the technical requirement of practicing transparency, the researcher suggests a hang tag including maximum detail information to the product that includes the name and contact information of a hypothetical manufacturer. This would allow a customer to verify environmentally and socially responsible methods practiced when the product was produced.

3.9 Consider waste-reducing design and clean manufacturing policy:

To reduce waste, the researcher suggested using eco-friendly dyeing methods such as reduced water dyeing or waterless dyeing method for the fabric processing. An on-product label with instructions to recycle, reuse of the product, thus reducing waste, should be added to the label. Clean production can be achieved by selecting manufacturers that make a commitment to the use of alternative or renewable energies. Maximum possible marker efficiency should be ensured during product development and manufacturing. Product printing, washing, and other finishing materials and processes should be selected focusing eco-sustainability priorities.

3.10 Raw materials

Selecting the eco-friendly raw materials is an important aspect for minimizing overall environmental impact throughout the product life cycles. Organic textile farming uses up to 60 % less water than conventional farming methods and were grown without cancer-producing pesticide and insecticide. Organic clothing is made from organic fibers grown without both exposures to toxins and irreversible environmental damage. Key factors of sustainable clothing are the fiber source and renewability without the use of agro-chemicals, hormones, and pesticides, along the entire manufacturing process from raw fiber to textile.

3.11 Consider natural and low-impact dyes and chemicals:

The designers, manufacturers, and merchandisers should introduce natural and environmentally low impact dyes in material processing for better product sustainability. Dyes and chemicals used for a product in wet processing, finishing, washing, and printing process should address both environmental concerns and economic realities with consumer demand for greener products.

4. CONCLUSIONS

The textile and apparel industries are notorious for their excessive contribution of waste and pollution to our environment. The unsustainable practice trends of poor designing, raw material selection, processing, manufacturing, and disposing of apparel products are the major contributing factors to this problem. The purpose of this study was to evaluate the material and manufacturing environmental sustainability of various branded T-shirts and find out various sustainable issues for greener apparel products. Based on eco-friendly test and Higg Index environmental assessment standard tool, the findings of this study could be helpful for many organizations and customers who attempt to produce and consume sustainable textiles and clothing products. Both the local and foreign branded products should be produced by strictly maintaining sustainability issues. Hence, this study is a little approach towards better sustainable practices and eventually would drive the textile and clothing designer, manufacturers, merchandisers, and consumers to introduce various sustainable issues in their works. Hence, this study leads to growing awareness and hopefully would direct them towards sustainable practices in the textile and clothing business for a greener world.

COMPETING INTERESTS

Both authors declare that they have no competing interest in relation to this study.

AUTHORS' CONTRIBUTIONS

MMI and MRK contributed to conception and design of the study. MMI carried out the experiments, analyzed the data as well as drafted the manuscript and MRK supervised the works. Both authors read and approved the final manuscript.

REFERENCES

1 Adams, CA, & Frost, GR. (2008). *Integrating sustainability reporting into management practices.*

2 Allwood, JM, Laursen, SE, de Rodriguez, CM, & Bocken, NMP. (2006). *Well dressed? The present and future sustainability of clothing and textiles in the United Kingdom.* Cambridge, UK: University of Cambridge, Institute of Manufacturing.

3 Challa L (2014) Impact of textiles and clothing industry on environment: approach towards eco- friendly textiles. http://www.fibre2fashion.com/. Accessed 28 June 2014.

4 Chen, HL, & Burns, LD. (2006). Environmental analysis of textile products. *Clothing and Textiles Research Journal, 24*(3), 248–261.

5 DEFRA (2010). Sustainable clothing action plan. https://www.gov.uk/government/uploads/system/uploads/attachment_data/file/69193/pb13206-clothing-action-plan-100216.pdf. Accessed 6 July 2014.

6 European Commission-Health and Consumers, Scientific Committees Opinion, see http://ec.europa.eu/health/scientific_

committees/environmental_risks/opinions/sctee/sct_out27_en.htm, Brussels, 18 January 1999.

7 Fletcher, K. (2008). *Sustainable fashion and textiles*. London, UK: Earthscan.

8 Fletcher, K, & Grose, L. (2012). *Fashion and sustainability: design for change*. London: Laurence King Publishing.

9 GB 18401, (2010) National general safety technical codes for textile products.

10 Goan, M. (1996). *An integrated approach to environmentally conscious design and manufacturing*. Doctoral dissertation, Virginia Polytechnic Institute and State niversity, Blacksburg, Virginia.

11 Goworek, H. (2011). Social and environmental sustainability in the fashion industry: a case study of a fair trade retailer. *Social Responsibility Journal, 7*(1), 74–86.

12 Islam, MM, Khan, MMR, & Elias, K. (2014). *American Journal of Engineering Research (AJER), 03*(12), 62–68.

13 IPP (2014) Integrated product policy (IPP). http://ec.europa.eu/environment/ipp/home.htm. Accessed 25 May 2014

14 Kadolph, SJ. (2010). *Textiles. Upper Saddle River*. NJ: Person Education Inc.

15 Laitala, K, & Boks, C. (2012). Sustainable clothing design: use matters. *Journal of Design Research, 10*(1), 121–139.

16 Laursen, S. E., Hansen, J., Knudsen, H. H., Wenzel, H., Larsen, H. F., and Kristensen, F. M. (2007). EDIPTEX: environmental assessment of textiles. Danish Environmental Protection Agency.

17 Lewis, VD, & Chem. (2006). The life of a piece of cloth: developing garments into a sustainable service system. *International Journal of Environmental, Cultural, Economic and Social Sustainability, 2*(1), 197–207.

18 Ljungberg, LY. (2007). Materials selection and design for development of sustainable products. *Materials in Engineering, 28*(2), 466–479.

19 Martin M (2013), Creating sustainable apparel value chains—a primer on industry transformation. http://www.ifc.org/. Accessed 25 Apr 2014

20 Md. Mazedul Islam and Md. Mashiur Rahman Khan, (2014). Environmental sustainability evaluation of apparel product: a case study on knitted T-shirt. Journal of Textiles. Volume: 2014, *643080*, pp: 1–6. http://dx.doi.org/10.1155/2014/643080.

21 Murray and Coody (2003), *Environmental impact of textile production.* (I. J. Ulasewicz, Ed.) New York: Fairchild Books, Inc.

22 Muthu, SS, Li, Y, Hu, JY, & Mok, PY. (2009). An exploratory eco-impact assessment of paper and plastic bags. *Journal of Fibre Bioengineering and informatics, 1*(4), 307–320.

23 Muthu, SS, Li, Y, Hu, JY, Mok, PY, Ding, X, Wang, X, & Weibang, C. (2010). Eco-impact of shopping bags: consumer attitude and government policies. *Journal of Sustainable Development, 3*, 71–83.

24 Muthu, SS, Li, Y, Hu, JY, & Mok, PY. (2012). Quantification of environmental impact and sustainability of textile fibres. *Ecological Indicator, 13*, 66–74.

25 Niinimäki, K, & Hassi, L. (2011). Emerging design strategies in sustainable production and consumption of textiles and clothing. *Journal of Cleaner Production, 19*(1), 1876–1883.

26 Nike Inc (2012c) Product design and materials. http://www.nikeresponsibility.com/report/content/chapter/materials. Accessed 2 Dec 2014.

27 Reichard R (2013) Textiles 2013: the turnaround continues. http://www.textileworld.com. Accessed 1 May 2014.

28 Reuben, A. (2013). The Higg index for sustainable apparel Environmental Performance Index. Retrieved December 12 from, http://epi.yale.edu/indicators/indicator-case-studies/reports/higg-index-.

29 Ruskino, C. (2007). Green manufacturing: an evaluation of environmentally sustainable manufacturing practices and their impact on competitive outcomes. *IEEE: Transactions on Engineering Management, 54*(3), 445–454.

30 SAC (2012i) SAC (2013) Material assessment, MSI information, MSI life cycle scope. http://www.apparelcoalition.org/msi/msi-information/msi-environmental-impact-areas.html. Accessed 8 Apr 2014.

31 SAC (2013a) Material sustainability index base material score. http://www.apparelcoalition.org/msi/. Accessed 5 May 2014.

32 SAC (2013c) Material assessment, MSI information, MSI environmental impact areas. http://www.apparelcoalition.org/msi/msi-information/msi-environmental-impact-areas.html. Accessed 2 May 2014.

33 Speer, J. (2005). Organic cotton: where, why and how. Apparel. *Journal of Sustainable Textiles Material, 46*(9), 29–34.

34 Steinberger, JK, Friot, D, Jolliet, O, & Erkman, S. (2009). A spatially explicit life cycle inventory of the global textile chain. *The International Journal of Life Cycle Assessment, 14*(5), 453–455.

35 USagain (2015). The environmental impact of cotton T-shirt. Retrieved March 15, 2015 from https://www.youtube.com/watch?v=g5rGm6veAhg.

36 The Higg Index. (2012). . Retrieved January 10, 2013 from http://www.apparelcoalition.org/higgindex.

37 Ulibarri S, (2011) sets standards for clothing and footwear. http://www.justmeans.com/blogs/sustainable-apparel-coalition-sets-standards-for-clothing-and-footwear. Accessed 20 May 2014.

38 Wolf MA, Pant R, Chomkhamsri K (2011), Towards life cycle sustainability management. http://link.springer.com/book/10.1007/978-94-007-1899-9. Accessed 1 March 2014.

CHAPTER 5

Surface Modification Methods for Improving the Dyeability of Textile Fabrics

Sheila Shahidi[1], Jakub Wiener[2] and Mahmood Ghoranneviss[3]

[1] Textile Department, faculty of Engineering, Arak Branch, Islamic Azad University, Arak, Iran

[2] Department of Textile Chemistry, Faculty of Textile, Technical University of Liberec, Liberec, Czech Republic

[3] Plasma Physics Research Center, Science and Research Branch, Islamic Azad University, Tehran, Iran

1. INTRODUCTION

Polymer and textiles have a vast number of advantages and attractiveness as a material. However, despite these advantageous, polymers have limitations. In general, special surface properties with regard to chemical composition, hydrophilicity, roughness, crystallinity, conductivity, lubricity, and cross-linking density are required for successful application of polymers in such wide fields as adhesion, membrane filtration, coatings, friction and wear, composites, microelectronic devices, thin-film technology and biomaterials, and so on. Unfortunately, polymers very often do not possess the surface properties needed for these applications. In fact, polymeric fibers that are mechanically strong, chemically stable, and easy to process usually will have inert surfaces both chemically and biologically. Vice versa, those polymers having active surfaces usually do not possess excellent mechanical properties which are critical for their successful application.

Due to this dilemma, surface modification of the polymeric fibers without changing the bulk properties has been a classical research topic for many years, and is still extensive studies as new applications of polymeric materials emerge, especially in the fields of biotechnology, bioengineering, and most recently in nanotechnology.

Modification is used to designate a deliberate change in composition or structure leading to an improvement in different type of fiber properties.

The challenge is, however, that there does not exist an ideal modification that eliminates all the negative properties and preserves all the positive properties of the fibers. This is why there are a great number of different single-purpose modifications. [1]

In spite of the great number of existing modification methods no consistent classification is available as yet. Some authors divide the methods into two groups depending on whether they involve changes in fiber composition (chemical modification) or changes in fiber structure (physical modification).

Surface modification of polymers has become an important research area in the plastic industry. Because polymers are inert materials and usually have a low surface energy, they often do not possess the surface properties needed to meet the demands of various applications. Advances in surface treatment have been made to rather chemical and physical properties of polymer surfaces without affecting bulk properties. Technologies such as surface modifications, which convert inexpensive materials into valuable finished goods, will become even more important in the future as material cost becomes a significant factor in determining the success of an industry. [2]

There are a few factors to consider when modifying a surface:

1. Thickness of the surface is crucial. Thin surface modifications are desirable, otherwise mechanical and functional properties of the material will be altered. This is more so when dealing with nanofibers as there is less bulk material present.

2. Sufficient atomic or molecular mobility must exist for surface changes to occur in reasonable periods of time. The driving force for the surface changes is the minimization of the interfacial energy.

3. Stability of the altered surface is essential, achieved by preventing any reversible reaction. This can be done by cross-linking and/or incorporating bulky groups to prevent surface structures from moving.

4. In some cases a transparent scaffold is desired, especially in optical sensors or ophthalmology; after surface treatment they should remain transparent. Any cloudiness introduced is of real concern.

5. Uniformity, reproducibility, stability, process control, speed, and reasonable cost should be considered in the overall process of surface modification. The ability to achieve uniform surface treatment of complex shapes and geometries can be essential for sensor and biomedical applications.

6. Precise control over functional groups. This is a challenging yet difficult scope. Many functional groups might bond to the surface such as hydroxyl, ether, carbonyl, carboxyl, and carbonate groups, instead of one desired functional group.

This review is focused on the application of recent methods for the modification of textiles using physical methods (corona discharge, plasma, laser, electron beam and neutron irradiations, Ion beam), chemical methods (ozone-gas treatment, surface grafting, enzymatic modification, sol-gel technique, micro-encapsulation method and treatment with different reagents). Nowadays, surface functionalization of synthetic fibers for various applications is considered as one of the best methods for modern textile finishing processes especially for improving the dyeability of fabrics. [3] Combination of physical technologies and nano-science enhances the durability of textile materials against washing, ultraviolet radiation, friction, abrasion, tension and fading. Textile fibers typically undergo a variety of pre-treatments before dyeing and printing is feasible. Nonetheless, these treatments still create undesirable process conditions which can result in increased waste production, unpleasant working conditions and higher energy consumption. Therefore reducing pollution in textile production is becoming of utmost importance for manufacturers worldwide. In coming years, the textile industry must implement sustainable technologies and develop environmentally safer methods for textiles processing to remain competitive. [4, 5]

1.1. Physical Methods for Modification of Textile Fabrics

1.1.1. Plasma Types and Applications

Plasma is by far the most common form of matter. Plasma in the stars and in the tenuous space between them makes up over 99% of the visible universe and perhaps most of that which is not visible.

The coupling of electromagnetic power into a process gas volume generates the plasma medium comprising a dynamic mix of ions, electrons, neutrons, photons, free radicals, meta-stable excited species and molecular and polymeric fragments, the system overall being at room temperature. This allows the surface functionalisation of fibres and textiles without affecting their bulk properties. These species move under electromagnetic fields, diffusion gradients, etc. on the textile substrates placed in or passed through the plasma. This enables a variety of generic surface processes including surface activation by bond breaking to create reactive sites, grafting of chemical moieties and functional groups, material volatilisation and removal (etching), dissociation of surface contaminants/layers (cleaning/ scouring) and deposition of conformal coatings. In all these processes a highly surface specific region of the material is given new, desirable properties without negatively affecting the bulk properties of the constituent fibres.

Plasmas are acknowledged to be uniquely effective surface engineering tools due to:

- Their unparalleled physical, chemical and thermal range, allowing the tailoring of surface properties to extraordinary precision.
- Their low temperature, thus avoiding sample destruction.

- Their non-equilibrium nature, offering new material and new research areas.
- Their dry, environmentally friendly nature. [5, 6]

1.1.2 Plasma reactors

Different types of power supply to generate the plasma are:

Low-frequency (LF, 50–450 kHz)

Radio-frequency (RF, 13.56 or 27.12 MHz)

Microwave (MW, 915 MHz or 2.45 GHz)

The power required ranges from 10 to 5000 watts, depending on the size of the reactor and the desired treatment.

1.1.3 Low-pressure plasmas

Low-pressure plasmas are a highly mature technology developed for the microelectronics industry. However, the requirements of microelectronics fabrication are not, in detail, compatible with textile processing, and many companies have developed technology of low pressure reactors to achieve an effective and economically viable batch functionalisation of fibrous products and flexible web materials.

A vacuum vessel is pumped down to a pressure in the range of 10^{-2} to 10^{-3} mbar with the use of high vacuum pumps. The gas which is then introduced in the vessel is ionised with the help of a high frequency generator.

The advantage of the low-pressure plasma method is that it is a well controlled and reproducible technique.

1.1.4 Atmospheric pressure plasmas

The most common forms of atmospheric pressure plasmas are described below.

1.1.5 Corona treatment

Corona discharge is characterised by bright filaments extending from a sharp, high-voltage electrode towards the substrate. Corona treatment is the longest established and most widely used plasma process; it has the advantage of operating at atmospheric pressure, the reagent gas usually being the ambient air. Corona systems do have, in principle, the manufacturing requirements of the textile industry (width, speed), but the type of plasma produced cannot achieve the desired spectrum of surface functionalisations in textiles and nonwovens. In particular, corona systems have an effect only in loose fibres and cannot penetrate deeply into yarn or woven fabric so that their effects on textiles are limited and short-lived. Essentially, the corona plasma type is too weak. Corona systems also rely upon very small interelectrode spacing (1 mm) and accurate web positioning, which are incompatible with 'thick' materials and rapid, uniform treatment.

1.1.6 Dielectric barrier discharge (silent discharge)

The dielectric barrier discharge is a broad class of plasma source that has an insulating (dielectric) cover over one or both of the electrodes and operates with high voltage power ranging from low frequency AC to 100 kHz. This results in non-thermal plasma and a multitude of random, numerous arcs form between the electrodes. However, these microdischarges are nonuniform and have potential to cause uneven treatment.

1.1.7 Glow discharge

Glow discharge is characterised as a uniform, homogeneous and stable discharge usually generated in helium or argon (and some in nitrogen). This is done, for example, by applying radio frequency voltage across two parallel-plate electrodes. Atmospheric Pressure Glow Discharge (APGD) offers an alternative homogeneous cold-plasma source, which has many of the benefits of the vacuum, cold-plasma method, while operating at atmospheric pressure. [5]

Cold plasmas can be used for various treatments such as: plasma polymerisation (gaseous monomers); grafting; deposition of polymers, chemicals and metal particles by suitable selection of gas and process parameters; plasma liquid deposition in vaporised form.

In general, reactions of gas plasmas with polymers can be classified as follows:

1. Surface reactions: Reactions between gas-phase species and surface species and reactions between surface species produce functional groups and cross-links, respectively, at the surface. Examples of these reactions include plasma treatment by argon, ammonia, carbon monoxide, carbon dioxide, fluorine, hydrogen, nitrogen dioxide, oxygen, and water.

2. Plasma polymerization: The formation of a thin film on the surface of polymer via polymerization of an organic monomer such as CH_4, C_2H_6, C_2F_4, or C_3F_6 in plasma. It involves reactions between gas-species, reactions between gas-phase species and surface species, and reactions between surface species.

3. Etching: Materials are removed from a polymer surface by physical etching and chemical reactions at the surface to form volatile products. Oxygen plasma and oxygen and fluorine-containing plasmas are frequently used for the etching of polymers. [6, 7]

Plasma treatment play very important role for improving the dyeing properties of textile fabrics. Some of these improvements are discussed as follow.

Surface Modification Of Polypropylene Non-Woven Fabrics By Atmospheric-Pressure Plasma Activation Followed By Acrylic Acid Grafting

Polypropylene (PP) non-woven fabrics have been activated by an atmospheric-pressure plasma treatment using surface dielectric barrier discharge in N_2 and ambient air. Subsequently, the plasma activated samples were grafted using catalyst-free water solution of acrylic acid. Surface properties of the activated and polyacrylic acid post-plasma grafted non-woven were characterized by scanning electron microscopy, Fourier transform infrared spectroscopy, electron spin resonance spectroscopy, surface energy and dyeability measurements.

The grafted non-woven exhibit improved water transport and dyeing properties.

The plasma activation in nitrogen plasma gas was more efficient than in air. Post-plasma surface grafting lead to a stable and homogeneous grafting of pAA onto PP non-woven fabrics, which made PP fabrics easily coloured by conventional water-soluble acid dye. Supposedly, peroxy radicals formed at a short ambient air exposure of the plasma activated fabrics were responsible for initiating the grafting process. Regarding the surface peroxy radicals generation, the nitrogen plasma gas was superior to ambient air and provided better grafting. [7-9]

One-Bath One-Dye Class Dyeing Of Pes/Cotton Blends After Corona And Chitosan Treatment

The feasibility of one-bath one-dye class dyeing of PES/cotton blend with direct (Direct Red 80) or reactive (Reactive Red 3) dye after pre-treatment with corona discharge (CD) and chitosan has been investigated by Ristic and his coworkers. It has been confirmed that corona discharge treatment enhances hydrophilicity of cotton and PES fibers due to surface modification of the material and formation of C-O, C=O, and COOH groups. Increased hydrophilicity of PES and cotton induced only slightly increased color intensity (K/S) with both dyes investigated. Nevertheless, subsequent treatment with biopolymer chitosan noticeably enhanced the color intensity obtained, especially with PES fiber. The dyeability improved proportionally with the concentration of chitosan treatment solutions. However, the highest values of color intensity were obtained for the PES/cotton fabric subjected to the combined CD and chitosan treatment, suggesting that CD pre-treatment enhances efficiency of chitosan application.

Satisfactory values of dye fastness and fixation degree of reactive dye were obtained. [10]

Surface And Bulk Cotton Fibre Modifications: Plasma And Cationization. Influence On Dyeing With Reactive Dye

In similar research work, the single or combined effects of corona air plasma and cationising with an epihalohydrin have been evaluated on the surface and dyeing properties of openwork twill cotton fabrics. Dyeing was performed with a hetero-bis functional reactive dye. Wetting properties of cotton fabrics were improved with a very short corona plasma treatment and a double side-effect was observed on the dyed fabric by contact angle analysis, because of the low penetration of the plasma on the fabric. Exhaustion of the dye and colour intensity of the cotton fabrics were increased due to the plasma treatment. This is well explained by the functionalisation of the surface with oxygenated moieties, without any significant alteration in surface topography of the fibres. Cationising of the cotton fabrics using an epihalohydrin as cationising agent increases the exhaustion of the dyestuff as high as 90%, and produces a dramatic improvement (80% increase) in the colour intensity (K/Scorr) on both sides of the fabrics. The improvement in colour intensity of the cationised cotton fabrics can be explained taking into account that the hydrolised reactive dye has high anionic character which can be bound to the cationic amine of the cationic agent on the cotton fabrics. It has been observed that plasma treatment previous to cationising increases the impregnation of the fabrics. [11]All theses possible effects are schematically represented in Figure 1.

The Impact Of Corona Modified Fibres' Chemical Changes On Wool Dyeing

Corona/plasma treatment is an environmentally friendly process applied to wool fabrics. The main contribution of the present work was to study the impact of Corona on dyeability of wool fibers. First, the different chemical aspects of a woven wool fabric's surface were determined using two different analytical skills (XPS and polyelectrolyte titration).The results show that, low-temperature plasma treatment has ability to change wool fibre morphology which could have an impact on sorption properties. fabrics were dyed with blue acid and blue metal-complex dyes, and dyeing behaviour were studied by means of on-line VIS spectrophotometry. Finally, dyed samples were colourimetricaly evaluated and colour differences were calculated. The results provided evidence that the overall carbon content was decreased while oxygen and nitrogen atoms were increased when using ionized air for fabric modification. It has also been noted that the amount of positive-charged functional groups in various pH ranges are higher for Corona-treated wool fabric in comparison with the untreated sample.

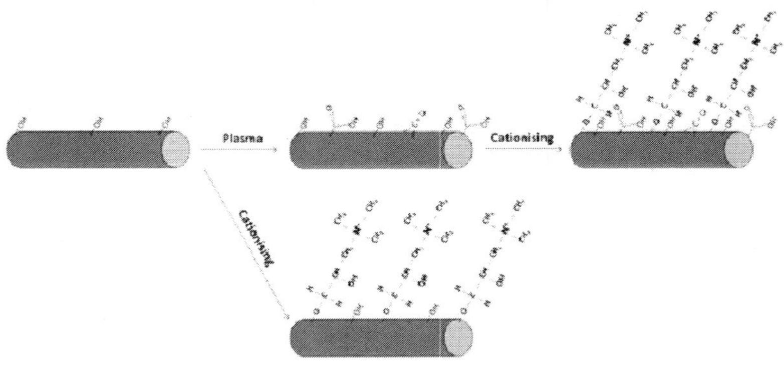

Figure 1. Effects of plasma and cationising processes in the surface functionalization of the cellulose constituting the cotton fibre

The surface performances of untreated and Corona-treated wool fabrics were studied both morphologically and chemically. Corona treatment is confirmed as inducing chemical and physical changes on the surface such as oxidizing/removing external fatty-acid monolayer, enlarging positively charged functional groups, creating new dyesites, and therefore, improving the exhaustion rate during dyeing.

Corona treatment applied in the pre-treatment stages of wool production can lead to an optimization of different dyeing procedures, implying lower dyeing temperatures and shorter dyeing time, achieving the same or even better colour exhaustion in comparison to conventional pre-treated wool fabric. For these reasons, the energy consumption can be reduced, thus also enhancing environmental protection. [12]

1.1.2. Surfaces Modification with Ionization Radiation

Ionization radiation such as high-energy electrons, X-rays, and gamma-rays can displace electrons from atoms and molecules, producing ions. It differs from other types of radiation such as infrared, visible, and UV in that it is highly energetic and delivers to the irradiated material a large amount of energy, much greater than that associated with chemical bonds. Common industrial ionization radiation sources are high-energy electrons (0.1-10 MeV) and cobalt-60 sources, and gamma radiation.

Electron beams from 0.1 to several mega electron volts are used for high doses and high speeds in various industrial processes, with penetration up to several millimeters for polymeric materials.

Surface modification by UV and IR lasers is useful in some specific applications. One key advantage of laser treatment is that the area to be treated can be very small and localized. Depending on the level of power chosen, ablation or chemical and physical changes can occur. Various chemical changes occur on photon-irradiated polymer surfaces. When PTFE was irradiated with ArF laser at high fluencies, defluorination and surface oxidation occurred. For polypropylene, formation of oxygen functional groups such as C-O and C=O groups was detected after UV laser irradiation in air and water, and in ozon. The treated surfaces were shown to have improved bond ability with an epoxy adhesive. The surface of poly(vinyl chloride) becomes electrically conductive after successive UV irradiation in chlorine and nitrogen and argon laser irradiation in air. These types of surface modification are very useful for increasing the dyeability of polymeric textile fabrics. [2] Some of irradiation dyeability modification is discussed below.

1.1.2.1. Effect Of Uv And Gamma Radiation On The Fabric Dyed With Natural Dyes

The irradiation of fabric is also another factor which affects the colour strength of the fabric. Previous studies show that UV irradiation adds value to colouration and also increases the dye uptake ability of the cotton fabrics through oxidation of surface fibers of cellulose. Gamma rays are ionizing radiations that interact with the material by colliding with the electrons in the shells of atoms. They lose their energy slowly in material being able to travel through significant distances before stopping. The free radicals formed are extremely reactive, and they will combine with the material in their vicinity. Upon irradiation the cross linking changes the crystal structure of the cellulose, which can add value in colouration process and causes photo modification of surface fibers. The irradiated modified fabrics can allow: more dye or pigment to become fixed, producing deeper shades, more rapid fixation of dyes at low temperature and increases wet ability of hydrophobic fibers to improve depth of shade in printing and dyeing. [13]

1.1.2.2. Surface Modification Of Meta-Aramid Films By Uv/Ozone Irradiation

Meta-aramid films surface was modified by UV/O_3 irradiation, and surface properties have been investigated by reflectance, ATR, ESCA, and surface energy. Upon UV/O_3 treatment, the surface roughness and the O1s/C1s atomic ratio obviously improved, resulting from the implantation of carbonyl and hydroxyl groups. The surface energy of the meta-aramid films increased substantially due to the significantly enhanced Lewis acid parameter, which promoted the acid-base interaction of the surfaces with increase in UV energy. Also meta-aramid films became hydrophilic as indicated by substantially decreased water contact angle.

The dyeability of aramid films to cationic dyes was significantly increased due to higher hydrophilic surface and strong electrostatic attraction between the cationic dyes and anionic dyeing sites of the meta aramid. [14]

1.1.2.3. Modification Of Polypropylene Fibers By Electron Beam Irradiation. I. Evaluation Of Dyeing Properties Using Cationic Dyes

The dyeing properties of hydrophobic polypropylene fibers using cationic dyes were investigated to improve dyeability by electron beam irradiation and sulfonic acid incorporation. The best dyeing result was obtained when polypropylene fibers incorporated by sulfonic acid group after electron beam irradiation were dyed with cationic dyes at alkaline conditions and 30~75 kGy irradiation ranges. In order to improve the dyeing properties of electronic beam irradiated polypropylene, sulfonic acid group which has good reactivity was introduced on the fiber. To incorporate sulfonic acid with the electronic beam irradiated (70.5 kGy) polypropylene fiber, the fibers were added and reacted to the solution of 1,4-dioxane and $ClSO_3H$ at 70 °C (Figure 2).

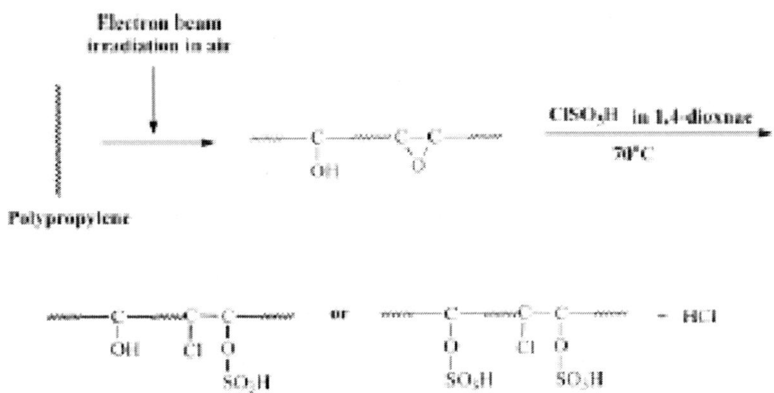

Figure 2. Introduction of sulfone groups on electron beam irradiated PP.

In order to make hydrophobic polypropylene fibers dyeable, it was shown that functional group such as carboxylate was formed on fiber substrates by electronic beam irradiation.

Concerning the pH and amount of absorbed electronic beam irradiation, the color strength increased as pH increased in alkaline conditions, and also increased as the absorbed dose increased to 30~75 kGy. As a result, it was confirmed that the pH of the dyebath and the amount and the range of the absorbed irradiation could be important variables for color strength but it seemed difficult to get deeper colors.

In the case of polypropylene fibers incorporated by sulfonic acid group to improve dyeability, the introduction of sulfonic acid group was confirmed by ESCA analysis and it was judged that such introduction has some advantages in color strength over only electronic beam irradiated fibers. Finally, the wash fastness of dyed fabrics using cationic dyes showed satisfactory ratings of 4~5 on both electronic beam irradiated fibers and sulfonic acid incorporated fibers. [15]

1.1.2.4. Uv Excimer Laser Modification On Polyamide Materials: Effect On The Dyeing Properties

Polyamide materials irradiated with 193 nm ArF Excimer laser developed ripple-like structures of micron size on the surface. (Figure 3)

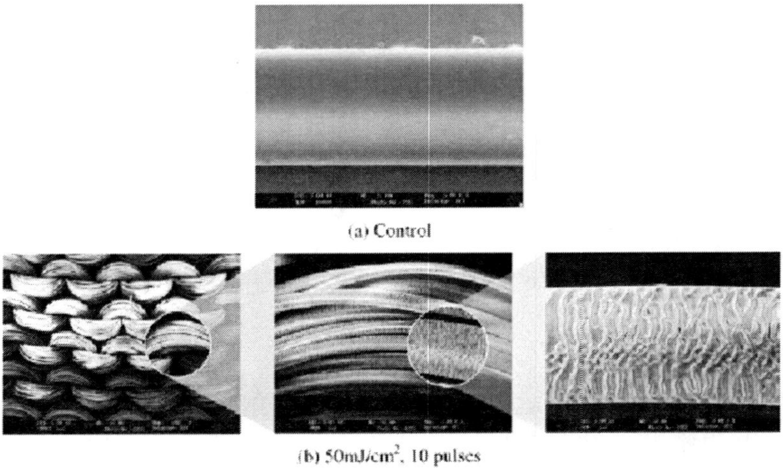

Figure 3. Morphological features of untreated and laser treated nylon 6 fabric

These structures are strictly perpendicular to the stress direction of the fiber. Dyeing results revealed that the dyeing properties of all dyes on polyamide fabrics changed remarkably after the treatment.

Results suggest that the change in coloration closely correspond with the ripple-like structures and the changes in chemical properties induced by laser treatment. It should be noted that the increased rate of exhaustion of acid dyes by laser treatment is not beneficial for dyeing polyamides since this will enhance the non-uniformity of the dye. While deeper dyeing is a greater advantage for disperse and reactive dyes since darker shades are obtainable using only the usual amounts of dyestuffs. To conclude, the excimer laser modification process has a high industrial potential, as it is an environmentally friendly dry process not involving any of the solvents required for a wet chemical process. After laser treatment, the dyeing properties of disperse and reactive dyes are improved and this provides an alternative choice for dyeing polyamide materials. (Figure 4) [16]

1.2. Chemical Modification Of Fabrics

Chemical treatment has been used in industry to treat large objects that would be difficult to treat by other commonly used industrial technique such as flame and corona-discharge treatments. Chemical etchants are used to convert smooth hydrophobic polymer surfaces to rough hydrophilic surfaces by dissolution of amorphous regions and surface oxidation. Chromic acid is the most widely used etchant for polyolefins and other polymers. [2]

Alkaline, acidic and solvents hydrolysis is another method to improve various physical and chemical properties of synthetic fibers. The alkaline hydrolysis of PET fibers is usually carried out with an aqueous alkaline solution, such as sodium hydroxide. In the alkaline hydrolysis process, PET undergoes a nucleophilic substitution. Chain scission of PET occurs, resulting in a considerable weight loss and the formation of hydroxyl and carboxylate end groups, which improves the handling, moisture absorption and dyeability of the fabric with enhanced softness.

There are different kinds of auxiliaries that are used to modify the surface properties of textile fabrics. Some of new chemical and their applications in textile industry are described bellow:

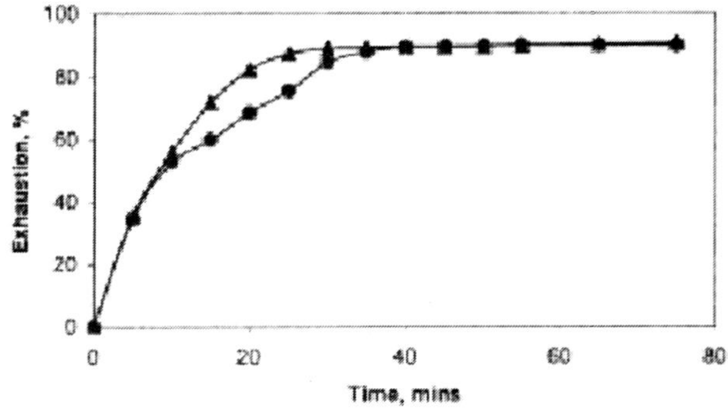

(a) Acid Dye

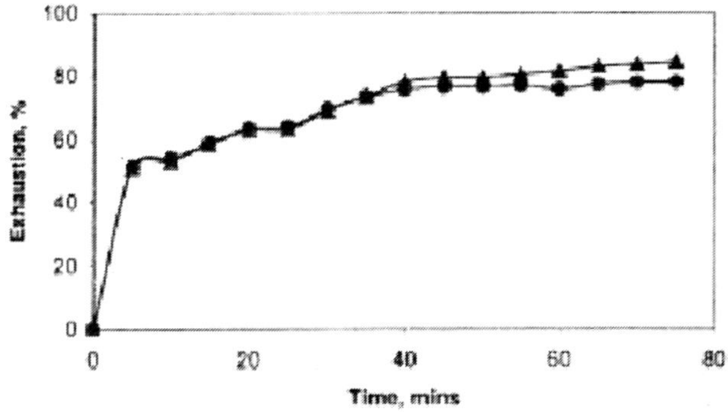

(b) Dispersive Dye

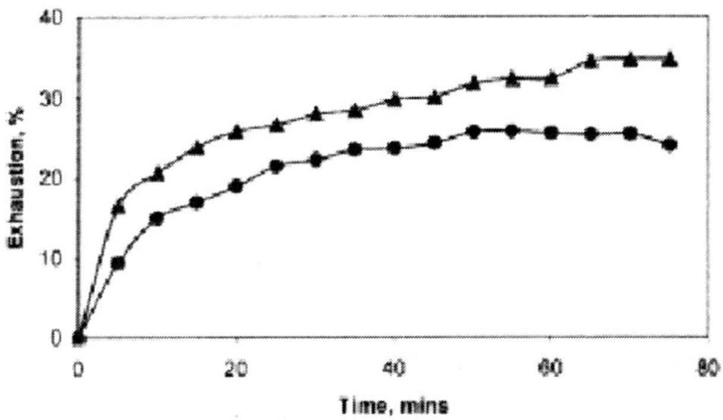

(c) Reactive Dye

Figure 4. Dye bath exhaustion study of laser treated nylon 6 fabrics; (●) Control (▲) Laser treated

1.2.1. Aminolysis

Several studies have assessed the effects of amine interaction with polyester. Early studies assessed the aminolysis of polyester as a means of examining fiber structure without regard to maintaining the integrity of the polymer.

The degradation effects on polyester of a monofunctional amine versus alkaline hydrolysis have been studied. These studies, which again involved high levels of fiber degradation, demonstrated that alkaline hydrolysis has a more substantial effect on fiber weight without extensive strength loss. In contrast, aminolysis had less effect on fiber weight but decreased fiber strength, indicative of a reaction within the polymer structure rather than simply at the surface. It was later demonstrated that bifunctional amine compounds could be reacted with the polymer with minimal loss in strength while generating amine groups at the fiber surface. The early stages of the reaction were largely confined to the fiber surface and the resulting fiber had modified wetting properties and improved adhesion with the matrix when used in composites. A recent paper has re-examined the interaction of untreated and alkali hydrolyzed polyester with a range of aliphatic diamines. 1,6-Hexanediamine, 2-methylpentamethylene diamine, 1,2-diaminocyclohexane, tetraethylenepentamine, and ethylene diamine were applied to untreated polyester. Ethylene diamine was also applied for a range of solution concentrations in toluene. The treatment generated amine groups on the fiber surface and was revealed by staining with anionic dyes under conditions in which the amine group was protonated. Unexpectedly, the reaction resulted in the simultaneous formation of carboxylic acid groups in a manner similar to alkaline hydrolysis, revealed by staining with Methylene Blue

(Figure 5). The reaction thus resulted in a bifunctional polyester surface. The ratio of amine and carboxylic acid groups differed with unhydrolyzed and hydrolyzed starting materials (Figure 5). Strength loss was somewhat greater than with alkaline hydrolysis. [17]

1.2.2. Sol Gel Pre Modification Of Fabrics

The sol-gel process is an excellent tool to obtain ordered hybrid organic-inorganic nanocomposites. The method involves the mixing of precursors in an aqueous or alcoholic medium. Precursors are molecules, which contain a central metal or semimetal atom, to which reactive alkoxy groups and/or organic groups are bonded. These reactive groups are subjected to an acidic or alkaline catalyzed hydrolysis and condensation reaction, thus forming a sol and subsequently a gel. Aging or drying step enables the production of powders, xerogels, aerogels, fibers, or coatings [18]. The latter procedure renders possible surface modification of textiles, thus imparting novel properties to the material.

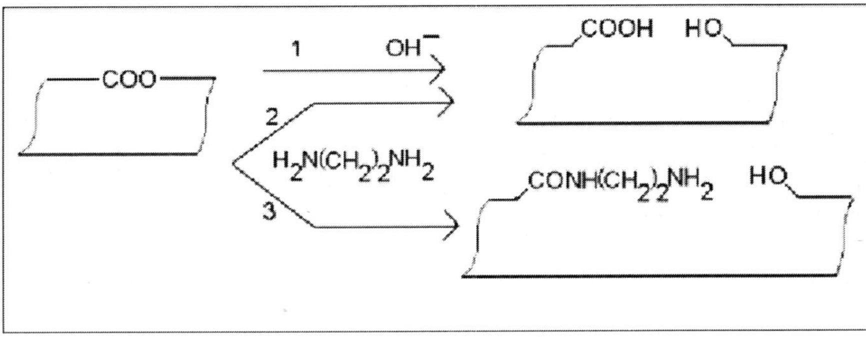

Figure 5. Hydrolysis and aminolysis of polyester

The sol-gel technology was also applied to influence the dyeing properties. Luo et al. [19] succeeded in improving the wash fastness of cotton materials that had been dyed with direct dyes using GPTMS and TEOS. Mahltig incorporated dyes into a silica matrix [20-22]. Du investigated the fixation properties and mechanism of direct dyes on silk, nylon 6 and cotton applying various organotrialkoxysilanes. They found that the most suitable precursors for silk and nylon 6 are amide- and vinyl-containing sols [14]. Yin also managed to enhance the fastness properties of cotton material that had been dyed with direct dyes [22].

1.2.2.1. Dyeing Treatment Of Sol-Gel Pre-Treated Cotton Fabrics

Schramm and his coworker investigated the impact of alkoxysilanes (TEOS, GPTMS, APTES, and TESP-SA) on the dyeing process of cotton substrates, which were dyed with 4 % owf C.I. Reactive Red 141 and 4

% owf C.I. Reactive Black 5. (Figure 6) For this purpose cotton samples were pre-treated with the alkoxysilanes and subsequently dyed. The results show that TESP-SA are lowering the L* values significantly, whereas TEOS, GPTMS, and APTES give rise to a moderate change of L*. The after-treatment of dyed cotton fabrics with alkoxysilane causes almost no effect with respect to the colorimetric data. The direct incorporation of the alkoxysilanes into the dyeing bath resulted in a reduction of the color properties, when APTES or TESP-SA was employed. The crease-proof finishing treatment caused an increase in the a* value of red-dyed samples. [23]

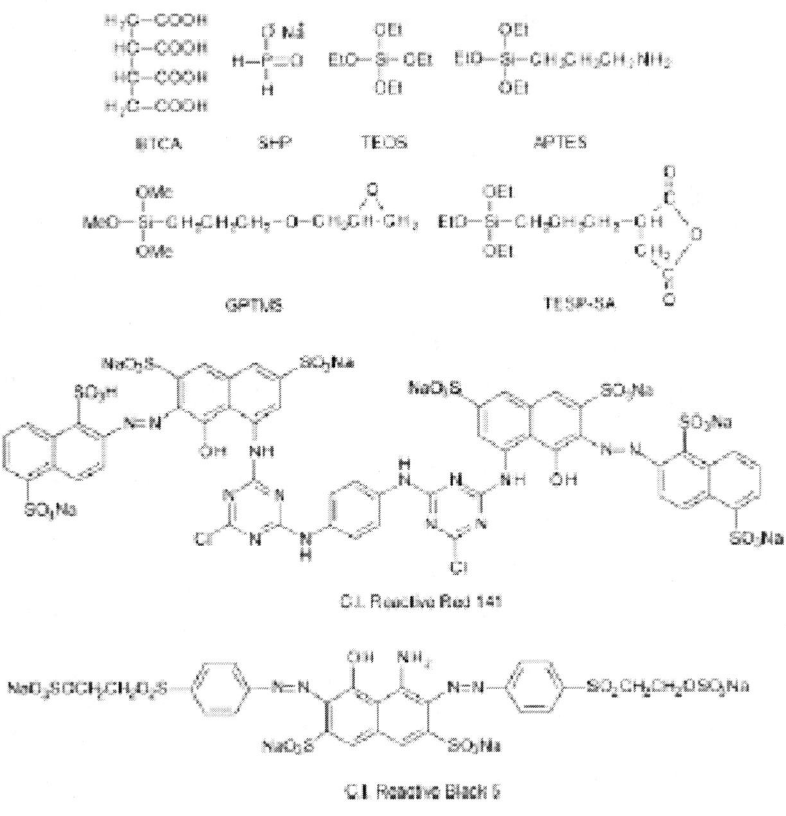

Figure 6. Chemical formula of the substances of interest.

1.2.2.2. An Evaluation Of The Dyeing Behavior Of Sol–Gel Silica Doped With Direct Dyes

Direct dyes are important dyes used on cellulose fiber directly. They can be applied in the same dyeing bath with other dyes. Moreover, the price is much lower than that of other dyes. However, several problems often occur in dyed fabric with direct dyes, such as lower washing and rubbing fastnesses. Some direct dyes perform rather poorly with respect to washing and rubbing fastnesses. They are mainly caused by the water-soluble groups, sulphonic and/ or carboxyl groups. Without an appropriate treatment, direct dyes bleed a little with each washing, lose their color on fabric and endanger other clothes washed in the same bath. At present, copper salt fixing reagent and cationic fixing reagent are widely used in textile industry to improve the fastness by cross-linking reactions between metal ion or formaldehyde and other molecules. But metal ions in fixing reagent such as cupric cation (Cu2+) will aggravate the difficulty of waste processing. Also some reagents and intermediates used in diazotization or crosslinking reactions such as formaldehyde, nitroaniline seriously affect the human health and the environment.

Moreover, the vividness after copper salt or diazotization finishing usually fades away, so that the application scope of direct dyes have been seriously limited in textile industry. A novel dyeing solution containing silica and direct dyes has been prepared by the sol–gel process. During this process, EtOH, tetraethoxysilane (TEOS), H2O and 3-glycidoxypropyltrimethoxysilane (GPTMS) were added in turn. The molar ratio of TEOS:H2O:EtOH was 1:5:8 and the concentration of GPTMS was 0.05 mol/L. Fabric was dyed at 90°C for 40 min. The concentration of NaCl added into the dyeing solution was 10 g/L. The dyed fabrics were baked at 150°C for 5 min. With this process, the results indicate that the K/S value is enhanced by more than 10%, the rubbing fastness and the washing change fastness are improved by one grade and the washing staining fastness is improved by half a grade. Not only is the K/S value enhanced from 9.3 to 11.5, but also the wet rubbing fastness and the washing change fastness are increased half a grade. Using a video microscope, a smoother fiber surface is observed. The calculated sol–gel weight gain is 4.6%. As a nonpolluting process, the sol–gel technology shortens the dyeing process and brings a better fixation property, meeting the needs of energy-saving and pollution-free processes.

1.2.3. Application Of Nanoparticles For Surface Modification Of Fibers

The dyeability of synthetic fibers depends on their physical and chemical structure. Dyeing process consist of three steps including the diffusion of dye through the aqueous dye bath on to the fiber, the adsorption of dye into the outer layer of the fiber and the diffusion of dye from the adsorbed surface into the fiber interior. It was shown by researchers that functional groups of PET and water molecules play a great role in this process. The terminal carboxylic and hydroxyl groups in PET chains interact with water molecules. This makes a swelled fiber resulting to increase the attraction of disperse dye by these functional groups of fiber.

The proportion of crystalline and amorphous regions of polymer is another factor influencing the dyeability. Researchers are concerned with the development and implementation of new techniques in order to fulfill improvement in dyeability of various polymers. Blending of polymeric fibers with nanoclays as inexpensive materials is still claimed as cost effective method to enhance dyeability. Up to now, only two research articles are focused on dyeing properties of polypropylene- and polyamide 6- layered clay incorporated nanocomposites prepared by melt compounding. Toshniwal et al. suggested that polypropylene fibers could be made dyeable with disperse dyes by addition of nanoclay particles in polymer matrix [25].

The previous study on dyeability of PET/clay nanocomposites stated the following type of interactions between the disperse dye and clay surfaces:

- Hydrogen bonding between OH groups of modified clays and the NH2 and CO groups of disperse dye molecules.

- Electrostatic bonding between the negatively charged oxygen atom of carbonyl groups in disperse dye molecule and positively charged nitrogen atom of quaternary ammonium salt in modified clays.

- Direct π interactions and van der Waals forces between methyl and ethyl groups of modified clays on one hand and methoxy group and benzene rings of disperse dye molecule on the other hand.

1.2.4. Application Of Cyclodextrins In Textile Dyeing

Cyclodextrins can be considered as a new class of auxiliary substances for the textile industry. Cyclodextrins can be used for textile application because of their natural origin and their biodegradability. Cyclodextrins play on important role in textile scientific research area and should play a significant role in the textile industry as well to remove or substitute various auxiliaries or to prepare textile materials containing molecular capsules which can immobilize perfumes, trap unpleasant smells, antimicrobial reagents and flame retardants. As cyclodextrins can incorporate different dyes into their cavity, they should be able to act as retarders in a dyeing process. Various auxiliary products are used in wet finishing processes, especially in dyeing and washing. One of the dyeing auxiliary products are leveling agent. Leveling agents help to achieve uniform dyeing by slowing down the dye exhaustion or by dispersing the dye taken by the fibre in a uniform way. They can be classified into two groups: agents having affinity to the dye and agents having affinity to the fibre. Agents having affinity to the dyes slow down the dyeing process by forming complexs with the dyes. The complex compound moves slower compared to the dye itself; at higher temperature the dye is released and it can be fixed to the fibre. Application of cyclodextrins as leveling agents having affinity to dyestuffs has been investigated in research work about dyeing of cellulose fibres with direct dyes by exhaust method where β-cyclodextrin was tested as a dye complexing agent. cyclodextrins as a dye retardant in the dyeing of PAN fibres with cationic dyes was studied; further it was reported that some azo disperse dyes formed inclusion complexes with cyclodextrins. [26]

1.2.5. Chitosan Applications in Textile Dyeing

Nowadays, the surface modification of textile fibres is considered as the best route to obtain modern textile treatments.

Among various available biopolymers, the polysaccharide chitosan (CHT) is highly recommendable, since it shows unique chemical and biological properties and its solubility in acidic solutions makes it easily available for industrial purposes. The polysaccharide-based cationic biopolymer chitosan is poly(1,4)-2-amino-2-deoxy-b-D-glucan, usually obtained by deacetylation of chitin that is widely present in the nature as a component of some fungi, exoskeleton of insects and marine invertebrates (crabs and shrimp). The chemistry of chitosan is similar to that of cellulose, but it reflects also the fact that the 2-hydroxyl group of the cellulose has been replaced with a primary aliphatic amino group. Among many other uses, it has been recently shown that chitosan improves the dye coverage of immature fibres in cotton dyeing and that it could be successfully used as a thickener and binder in pigment printing of cotton.

Also Gupta et al showed that, chitosan treated cotton has better dyeability with direct and reactive dyes and treatment with modified chitosan makes it possible to dye cotton in bright shades with cationic dyes having high wash fastness. Treated samples showed good antimicrobial activity against Escherichia coli and Staphylococcus aureus at 0.1% concentration as well as improved wrinkle recovery. [28]

In similar research work, Jocic and his coworkers assessed the interactions that could occur during dyeing of the chitosan treated wool fibres, by measuring the absorbance values of the solutions containing dye and chitosan. It has been shown that there is a 1:1 stoichiometry between protonated amino groups and sulfonate acid groups on the dye ions in low concentrated chitosan solutions. This interaction forms an insoluble chitosan/dye product. With the excess of chitosan in the solution, the dye can be distributed between the different chitosan molecules and the soluble chitosan/dye products remain in the solution. It is suggested that the mechanism of the interaction involves the possibility of adsorbed dye molecules to be desorbed and redistributed between other components present in the system, depending on system parameters (pH, temperature and electrolyte presence). This fact is important in explanation of dyeing behaviour of chitosan treated wool and enables the assessment of the mechanism of dyeing of accordingly modified textile fibres [29].

Also Kitkulnumchai et al, showed that, the KIO4 oxidation of cellulose fabrics created more aldehydic groups on the fabric surface and the following reductive ammination with chitosan afforded stable C–N bonds between cellulose and chitosan chains. The attachment of chitosan to the fabric considerably improved dye uptake of mono chlorotriazine and vinyl sulfone reactive dyes resulting in greater dye exhaustion and the color yield (K/S).

The enhanced dyeability of the modified fabric is likely resulted from the reduction of the coulombic repulsion between the fabric surface and the anionic dye molecules in the presence of the positively charged chitosan on the surface. The oxidation step can cause some drop in the fabric strength. This oxidation reductive amination with chitosan is thus a convenient method for modifying the surface activity of cellulose fabrics whereas the fabric strength is not of the great priority.[30]

In another research, Cotton fabrics have been successfully dyed by green tea extract upon chitosan mordanting, and UV protection property of the chitosan mordanted green tea dyed cotton was increased.

The following conclusions have been made from this study;

1. Chitosan mordanting can effectively increase the ΔE and the K/S, that is, the dyeing efficiency of green tea dyeing onto cotton fabrics. As chitosan concentration increased, the ΔE and the K/S of cotton fabrics by green tea extract increased gradually.

2. Chitosan mordanting can effectively increase the UV protection property of both UV-A and UV-B of green tea dyed cotton fabrics. Chitosan mordanted undyed cotton and chitosan unmordanted dyed cotton did not show an increase in UV protection property. Therefore, it can be assumed that chitosan increased the uptake of active moiety, catechin, in green tea, which would be responsible for the UV protection and subsequently increased UV protection property of the chitosan mordanted green tea dyed fabric.

3. As chitosan mordanting concentration increased, UV protection property increased in both UV-A and UV-B.

Around 7% UV protection increase from control was observed upon chitosan mordanting, which is similar value of the green tea dyed cellulose fabric using a specific metal mordant. Therefore, it can be concluded that green tea dyeing can be used in developing UV protective cotton textiles, and the chitosan mordanting process would be necessary in green tea dyeing of cotton to increase not only the dyeing efficiency but also the UV protection property of cotton fabrics. [31]

REFERENCES

1. J. Militky, J. Vanicek, J.Krystufek, V. Hartych, Modified Polyester Fibers, Elsevier, 1991

2. Chan, Polymer Surface Modification and Characterization, Hanser Publisher, Munich Vienna New York, 1994

3. Renuga Gopal, Ma Zuwei, Satinderpal Kaur, and Seeram Ramakrishna Surface Modification and Application of Functionalized Polymer Nanofibers, Functionalized Polymer Nanofibers, 75

4. Mazeyar Parvinzadeh Gashti, Julie Willoughby and Pramod Agrawal, Surface and Bulk Modification of Synthetic Textiles to Improve Dyeability, textile dyeing, Intech, 2011 with F. solani pisi cutinase and pectate lyase. Enzyme & Microbial Technology, Vol. 42, No. 6, (May 2006), pp. 473-482, ISSN 0141-0229

5. R.Shishoo, Plasma Technologies for Textiles, Woodhead Publishing for textiles, Cambridge England, 2006

6. T.Oktem, Modification of polyester and polyamide fabrics by different in situ plasma polymerization methods, Turk J Chem, 24 (2000)-275-285.

7. Peter. P, Surface modification of fabrics using one atmosphere glow discharge plasma to improve fabric wettabilitty, Textile Research Journal, 67(5), 1997, 359-396.

8. L'. ˇCern´akov´a, D. Kov´aˇcik, A. Zahoranov´a M. ˇCern´ak, and M. Maz´ur, Surface Modification of Polypropylene Non-Woven Fabrics by Atmospheric-Pressure Plasma Activation Followed by Acrylic Acid Grafting Plasma Chemistry and Plasma Processing, Vol. 25, No. 4, August 2005

9. Nebojša Risti , Petar Jovan i , Cristina Canal, and Dragan Joci, One-bath One-dye Class Dyeing of PES/Cotton Blends after Corona and Chitosan Treatment , Fibers and Polymers 2009, Vol.10, No.4, 466-475

10. Alejandro Patin˜o Cristina Canal Cristina Rodrı´guez Gabriel Caballero Antonio Navarro Jose´ Ma Canal, Surface and bulk cotton

fibre modifications: plasma and cationization. Influence on dyeing with reactive dye Cellulose (2011) 18:1073–1083

11. Darinka Fakin, Alenka Ojstrsˇek, Sonja Cˇ elan Benkovicˇ The impact of corona modified fibres' chemical changes on wool dyeing, journal of materials processing technology 2 0 9 (2 0 0 9) 584–589
12. Javed, I., Bhatti, I.A., Adeel, S., 2008.Effect of UV radiation on dyeing of cotton fabric with extracts of henna leaves. Indian Journal of fiber and textile research, 33,157-162.
13. Eun-Min Kim and Jinho Jang , Surface Modification of Meta-aramid Films by UV/ozone Irradiation Fibers and Polymers 2010, Vol.11, No.5, 677-682
14. Hongje Kim and Jin-Seok Bae, Modification of Polypropylene Fibers by Electron Beam Irradiation. I.Evaluation of Dyeing Properties Using Cationic Dyes Fibers and Polymers 2009, Vol.10, No.3, 320-324
15. Joanne Yip Kwong Chan Kwan Moon Sin Kai Shui Lau, UV Excimer laser modification on polyamide materials: effect on the dyeing properties, Mat Res Innovat (2002) 6:73–78
16. Martin Bide, Matthew Phaneuf, Philip Brown, Geraldine McGonigle, and Frank LoGerfo, MODIFICATION OF POLYESTER FOR MEDICAL USES, Modified Fibers with Medical and Specialty Applications, 91–124, 2006
17. C. J. Brinker and G. W. Scherer, "Sol-gel Science, The Physics and Chemistry of Sol-gel Processing", Academic Press, San Diego, 1990.
18. M. Luo, X. Zhang, and S. Chen, Color. Technol., 119, 297 (2003).
19. B. Mahltig, H. Böttcher, D. Knittel, and E. Schollmeyer, Text. Res. J., 74, 521 (2004).
20. J. Du, Z. Li, and S. Chen, Color. Technol., 121, 29 (2005).
21. Y. Yin, C. Wang, and C. Wang, J. Sol. Gel. Sci. Technol., 48, 308 (2008).
22. Christian Schramm and Beate Rinderer, Dyeing and DP Treatment of Sol-gel Pre-treated Cotton FabricsFibers and Polymers 2011, Vol.12, No.2, 226-232

23. Yunjie Yin , Chaoxia Wang , Chunying Wang, An evaluation of the dyeing behavior of sol–gel silica doped with direct dyes , J Sol-Gel Sci Technol (2008) 48:308–314

24. Toshniwal, L., Fan, Q., Ugbolue, S.C., (2007) Dyeable polypropylene fibers via nanotechnology, Journal of Applied Polymer Science, 1, pp. 706–711.

25. Bojana Voncina, Application of Cyclodextrins in Textile Dyeing, Textile dyeing, Intech

26. Dragan Jocica,*, Susana Vı´lchezb, Tatjana Topalovicc, Antonio Navarroa, Petar Jovancicc, Maria Rosa Julia`b, Pilar Errab, Chitosan/acid dye interactions in wool dyeing system Carbohydrate Polymers 60 (2005) 51–59

27. Deepti Gupta a,*, Adane Haile, Multifunctional properties of cotton fabric treated with chitosan and carboxymethyl chitosan, Carbohydrate Polymers 69 (2007) 164–171

28. Dragan Jocic, Susana Vı´lchez, Tatjana Topalovic, Antonio Navarro, Petar Jovancic, Maria Rosa Julia`, Pilar Erra, Chitosan/acid dye interactions in wool dyeing system, Carbohydrate Polymers 60 (2005) 51–59

29. Yupaporn Kitkulnumchai, Anawat Ajavakom, Mongkol Sukwattanasinitt, Treatment of oxidized cellulose fabric with chitosan and its surface activity towards anionic reactive dyes, Cellulose (2008) 15:599–608

30. Sin-hee Kim*, Dyeing Characteristics and UV Protection Property of Green TeaDyed Cotton Fabrics–Focusing on the Effect of Chitosan Mordanting Condition–Fibers and Polymers 2006, Vol.7, No.3, 255-261

CHAPTER 6

Cyclodextrins in Textile Finishing

Bojana Voncina[1] and Vera Vivod[1]

[1] Faculty of Mechanical Engineering, University of Maribor, Maribor, Slovenia

1. INTRODUCTION

Chemical finishing is crucial for giving textiles new functionalities and making them appropriate for special applications, such as antimicrobial resistance, flame retardancy and others. Textile finishing is also an important process as it improves appearance, performance or hand.

Cyclodextrins can act as hosts and form inclusion compounds with various small molecules. Such complexes can be formed in solutions, in a solid state as well as when cyclodextrins are linked to various surfaces where they can act as permanent or temporary hosts for small molecules that provide certain desirable attributes. This characteristic makes cyclodextrin a promising reagent in textile finishing.

2. CYCLODEXTRINS

Cyclodextrins(CDs) are relevant molecules used in different applications and industries from pharmacology, cosmetics, textiles, filtration to pesticide formulations. They comprise a family of three well-known industrially produced substances. The practically important, industrially produced CDs are the α-, β-, and γ-CDs. There are some seldom used cyclic oligosaccharides as well but because of their cost are not applicable to industrial applications [1]. CDs are cyclic oligomers of a-D-glucopyranose that can be produced with the transformation of starch by certain bacterias such as Bacillus macerans [2, 3].

The preparation process of CDs consists of four principal phases: (i) culturing of the microorganism that produces the cyclodextrin glucosyl transferase enzyme (CGT-ase); (ii) separation, concentration and purification of the enzyme from the fermentation medium; (iii) enzymatical conversion of prehydrolyzed starch in mixture of cyclic and acyclic dextrins; and (iv) separation of CDs from the mixture, their purification and crystallization. CGT-ase enzymes degrade the starch and starts intramolecular reactions without the water participation. In the process, cyclic (CDs) and acyclic dextrins are originated, which are oligosaccharides of intermediate size. CDs are formed by the link between units of glucopyranose. The union is made through glycosidic oxygen bridges by α-(1,4) bonds. The purification of α- and γ-CDs increases the cost of production considerably, so that 97% of the CDs used in the market are β-CDs [2].

CDs ring structures act as hosts and form inclusion compounds with various small molecules. Such complexes can be formed in solutions, in a solid state as well as when cyclodextrins are linked to various surfaces. In all forms they can act as permanent or temporary hosts to small molecules that provide certain desirable attributes.

In the textile field CDs may have many applications such as: absorption of unpleasant odours; they can complex and release fragrances, "skin-care-active" and bioactive substances. Further, various textile materials treated with cyclodextrins could be used as selective filters for adsorption of small pollutants from waste water [4].

After the discovery of CDs scientists considered them poisonous substances and their ability for complexes formation was only considered a scientific curiosity. Later on, research on CDs proved that they are not only non-toxic but they can be helpful for protecting flavours, vitamins and natural colours [2]. CDs already play a significant role in the textile industry and can be used in dyeing, surface modification, encapsulation, washing, and preparation of polymers and in fibre spinning.

Since year 2000, β-CD has been introduced as a food additive in Germany. With respect to OECD experiments, this compound has shown no allergic impact. In the USA α-, β- and γ-CDs have obtained the GRAS list (FDA list of food additives that are 'generally recognized as safe') status and can be commercialized as such. In Japan α-, β- and γ-CDs are recognized as natural products and their commercialization in the food sector is restricted only by considerations of purity. In Australia and New Zealand α- and γ-CDs are classified as Novel Foods from 2003 and 2004, respectively. The recommendation of Joint FAO/WHO Expert Committee on Food Additives (JECFA) for a maximum level of β-CDs in foods is 5 mg/kg per day. For α- and γ-CDs no Acceptable Daily Intake (ADI) was defined because of their favourable toxicological profiles [2, 5].

Natural cyclodextrins and their hydrophilic derivatives are only able to permeate lipophilic biological membranes, such as the eye cornea, with considerable difficulty. All toxicity studies have demonstrated that orally administered cyclodextrins are practically non-toxic, due to lack of absorption from the gastrointestinal tract. The main properties of β-CD, the most important cyclodextrin in textile application are: less irritating than α-CD after i.m. injection, binds cholesterol, small amount (1-2%) is adsorbed in the upper intestinal tract, no metabolism in the upper intestinal tract, metabolised by bacteria in caecum and colon, LD50 oral rat > 5000 mg/kg, LD50 i.v., rat: between 450-790 mg/kg, however, application of high doses may be harmful and is not recommended [6, 7].

2.1. Structure of Cyclodextrins

The three mayor CDs are crystalline, homogeneous, and non-hygroscopic substances, which are torus-shaped oligosaccharides [1, 8-10], composed in the more common forms of six to eight (α-1,4)-linked α-D-glucopyranose units (α-, β- and γ-cyclodextrin), Figure 1 schematically present the main three cyclodextrins [11]. They are of circular and conical conformation, where the height is about 800 pm. The inner diameter of the cavity varies from 500 to 800 pm.

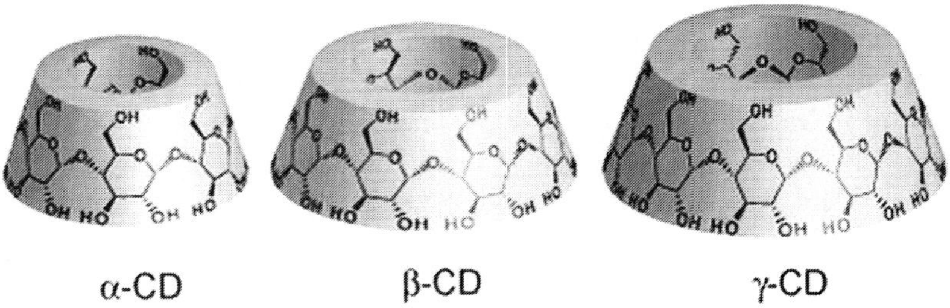

Figure 1. Schematic presentation of the main three CDs [**12**].

All glucopyranose units in the torus-like ring possess the thermodynamically favoured chair conformation because all substituents are in the equatorial position. As a consequence of the 4C1 conformation of the glucopyranose units, all secondary hydroxyl groups are situated on the lager side of the ring, whereas all the primary ones are placed on the narrower side of the ring. Hydroxyl groups on the outside of the CDs ensure good water solubility. The cavity is lined with hydrogen atoms of C3, by the glycosidic oxygen bridges and hydrogen atoms of C5. The nonbonding electron pairs of the glycosidic oxygen bridges are directed toward the inside of the cavity producing a high electron density and because of this the inner side of the cavity has some Lewis base characteristics. The C-2-OH group of one glucopyranoside unit can form a hydrogen bond with the C-3-OH group of the adjacent glucopyranose unit. In the CD molecule, a complete secondary belt is formed by these hydrogen bonds, therefore the β-CD has a rather rigid structure. Because of this arrangement, the interior of the toroids is not hydrophobic but considerably less hydrophilic than the aqueous environment and thus able to host other hydrophobic molecules. CDs behave more or less like rigid compounds with two degrees of freedom, rotation at the glucosidic links C4-O4 and C1-O4 and rotations at the O6 primary hydroxyl groups at the C5-C6 band. The intramolecular hydrogen bond formation is probably the explanation for the observation that β-CD has the lowest water solubility of all CDs.

The hydrogen-bond belt is incomplete in the α-CD molecule, because one glucopyranose unit is in a distorted position. Consequently, instead of the six possible H-bonds, only four can be established fully. γ-CD is a noncoplanar with more flexible structure; therefore, it is the most soluble of the three CDs. Figure 2 shows a sketch of the characteristic structural features of CDs. On the side where the secondary hydroxyl groups are situated, the diameter of the cavity is larger than on the side with the primary hydroxyls, since free rotation of the primary hydroxyls reduces the effective diameter of the cavity [13, 14].

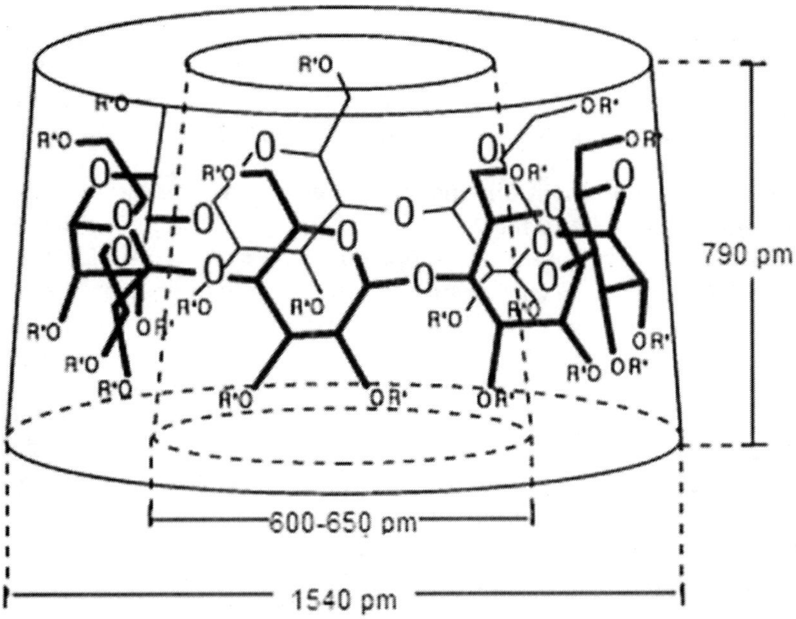

Figure 2. Schematic presentations of characteristic structural features of CDs.

By far, β-cyclodextrin (β-CD), with 7 sugar units, has been commercially the most attractive (more than 95% consumed) due to its simple synthesis, availability and price.

A β-CD molecule has a molecular weight of 1135 and a height of 750–800 pm. The internal diameter of the molecule's hole is between 600 and 680 pm, and the external diameter is 1530 pm[1, 15]. The volume of the hole is 260–265 Å^3, and the dissolution is 1.85 g/100 mL of water. The cavity is hydrophobic; the external section is hydrophilic in nature. β-CD is stable in alkali solutions and it is sensitive to acid hydrolysis [16].

2.2. Properties of Cyclodextrins

The most notable feature of CDs is their ability to form solid inclusion complexes ("host–guest" complexes) with a very wide range of solid, liquid and gaseous compounds by a molecular complexation. The phenomenon of CD inclusion compound formation is a complicated process involving many factors playing an important role.

Complex formation is a dimensional fit between host cavity and guest molecule. The lipophilic cavity of CD molecules provides a microenvironment into which appropriately sized non-polar moieties can enter to form inclusion complexes. No covalent bonds are broken or formed during formation of the inclusion complex.

According to some authors [11, 17] hydrophobic interactions are the main driving forces for CD-based host-guest compounds. Other requirements such as steric hindrance and relation between sizes of host and guest cavities are also important. This is illustrated by the fact that not only hydrophobic interaction will lead to an association between a guest molecule and a CD but ionic solutes, such as non-associated inorganic salts, can also be involved in these complexes.

Some researchers [7] claim that the main driving force of complex formation is the release of enthalpy-rich water molecules from the cavity. The water molecules located inside the cavity cannot satisfy their hydrogen bonding potentials and therefore are of higher enthalpy. The energy of the system is lowered when these enthalpy–rich water molecules are replaced with suitable guest molecules which are less polar than water. In an aqueous solution, the slightly apolar CD cavity is occupied by water molecules which are energetically unfavoured, and therefore can be readily substituted by appropriate "guest molecules" which are less polar than water. This apolar–apolar association decreases the CD ring strain resulting in a more stable lower energy state. The dissolved CD is the "host" molecule, and the "driving force" of the complex formation is the substitution of the high-enthalpy water molecules by an appropriate "guest" molecule.

The binding of guest molecules within the host CD is not fixed or permanent but rather is a dynamic equilibrium. Binding strength depends on how well the 'host–guest' complex fits together and on specific local interactions between surface atoms. Complexes can be formed either in solution or in the crystalline state and water is typically the solvent of choice. Inclusion complexation can be accomplished in a co-solvent system and in the presence of any non-aqueous solvent [7]. Generally, one guest molecule is included in one CD molecule, although in the case of some low molecular weight molecules, more than one guest molecule may fit into the cavity, and in the case of some high molecular weight molecules, more than one CD molecules may bind to the guest. In principle, only a portion of the molecule must fit into the cavity to form a complex. CD inclusion is a stoicmetric molecular phenomenon in which usually only one guest molecule interacts with the cavity of the CD molecules to become entrapped. 1:1 complex is the simplest and most frequent case. However, 2:1, 1:2, 2:2, or even more complicated associations, and higher order equilibrium exist almost always simultaneously.

Inclusion in CDs has a profound effect on the physicochemical properties of guest molecules as they are temporarily included within the host cavity.

These properties are:

- solubility enhancement of highly insoluble guests,
- stabilisation of labile guests against the degradative effects of oxidation, visible or UV light and heat,
- control of volatility and sublimation,
- physical isolation of incompatible compounds,
- chromatographic separations,
- taste modification by masking of flavours, unpleasant odours,
- controlled release of drugs and flavours,
- removal of dyes and auxiliaries from dyeing effluents,

- retarding effect in dyeing and finishing,
- protection of dyes from undesired aggregation and adsorption.

2.3. Application of Cyclodextrins

Complexes can be formed in solutions, in the solid state, as well as when CDs are linked to a solid surface where they can act as permanent or temporary hosts to those small molecules that provide certain desirable attributes such as adsorption of dyestuff molecules, fragrances or antimicrobial agents. This "molecular encapsulation" is already widely utilized in many industrial products, technologies, and analytical methods [7, 18].

Due to the relatively non-polar character of the cavity in comparison to the polar exterior, CD can form inclusion complexes with a wide variety of guest molecules, such as drugs [19, 20, 21], ionic and non-ionic surfactants [23, 24, 25], dyes [26, 27] and polymers [28], etc. The use of CDs has increased annually around 20–30%, of which 80-90% was in food products [29]. In the pharmaceutical industry, CDs and their derivatives have been used either for complexation of drugs or as auxiliary additives such as solubilizers, diluents, or ingredients for improving of drugs physical and chemical properties, or to enhance the bioavailability of poorly soluble moieties [30]. In the chemical industry, CDs and their derivatives are used as catalysts to improve the selectivity of reactions, as well as for the separation and purification of industrial-scale products [31]. In the food, cosmetics, toiletry, and tobacco industries, CDs have been widely used either for stabilization of flavours and fragrances or for the elimination of undesired tastes, microbiological contaminations, and other undesired compounds [7]. For the last 30 years, the use of CDs and their derivatives in the textile domains has captivated a lot of attention. Many of the papers and patents report the use of CDs for finishing and dyeing processes. For instance, they discuss the capture of unpleasant smells due to perspiration, or how to do the controlled release of perfumes, insecticides and antibacterial agents [4, 32-45].

3. CYCLODEXTRINS IN TEXTILES

In the textile field CDs may have many applications such as: they can absorb unpleasant odours; they can complex and release fragrances or "skin-care-active" substances like vitamins, caffeine and menthol as well as bioactive substances such as biocides and insecticides. Further, various textile materials treated with CDs could be used as selective filters for adsorption of small pollutants from waste waters - "preparation of textile nanosponges".

3.1. Cyclodextrins in Textile Finishing

One of the new concepts for modification of textile substrates is based on the permanent fixation of supramolecular compounds on the material's surface and, thus, imparts new functionality to the fabric. [46]. One of the most promising supramolecular moieties applied to textiles are CDs. Covalent bonding of CDs onto textile fibres was firstly patented in 1980 by Szejtli [47]. He and co-workers reported to bond CD via crosslinking reagent epichlorhydrin onto alkali-swollen cellulose fibres. They found out that CD covalently bonded to cellulose retained the ability to form complexes; when cellulose was treated with a drug it complexed with CD. The complexed drug was upon the contact with the skin released; cellulose textile substrate containing covalently bound β-CD was treated with solution of iodine, potassium iodide and methanol as a solvent to prepare a medical bandage [48].

Szejtli published [4] a very extensive review about CDs in the textile industry. In his review he divided the application of CDs in textile sectors in the following areas: binding of CDs to fibre surfaces, CDs in textile dyeing, in textile finishing, CDs and detergents and miscellaneous applications of CDs in textile industry and textile care. Due to the fact that application of CDs in textile dyeing processes was extensively reported in a chapter of the book edited by Hauser [18], we will emphasis in current publication about the application of CDs in textile finishing.

3.1.1. Binding of Cyclodextrins to Fibre Surfaces

The attachment of CD molecules on textile substrate provides hosting cavities that can include a large variety of guest molecules for specific functionality. There are two possible approaches to bond CDs onto textile fibres such as chemical bonding of modified CDs on the fibre surfaces or to use bifunctional reagents to link CDs covalently on fibre surfaces.

The most promising approach to bond modified CDs onto textile fibres is the modification of CDs with trichlorotriazines to prepare monochlorotriazinyl-cyclodextrin (CD-MCT) [49, 50]. Analogues to reactive dyes the CD-MCT can be fixed to the fabric by well-known methods and with common equipment. CD-MCT can be applied to fibre surfaces by an exhaustion method or by thermofixation. Moldenhauer with co-workers found out [51] that the fixation was the best when textile substrate was cotton. Mixed fibre materials like cotton/polyurethane or cotton/polyamide can be finished with β-CD-MCT in good yields. Ibrahim et.al [52] reported the improvement of UV protective properties of cotton/wool and viscose/wool blends via incorporating of reactive β-CD-MCT in the easy care finishing formulations, followed by subsequent treatment with copper-acetate or post-dyeing with different classes of dyestuffs (acid, basic, direct and reactive). They found out that post-dyeing of the prefinished textile blends results in a significant increase in the UPF (UV-protection factor) values as a direct consequence of a remarkable reduction in UV radiation transmission through the plain weave fabric. β-CD modified with monochlorotriazine was applied to the cotton fabrics for entrapping of sandalwood oil as an aroma-finishing agent by Sricharussin [53]. The Fourier transform infrared, tensile stress tests and gas chromatography-mass spectroscopy measurements were used to investigate the effects of the treatment. It was found that β-CD-MCT can be fixed to cotton fabrics with the pad-dry-cure method at high temperature. No loss of tensile strength of the treated fabrics was reported. The fragrance disappeared from untreated cotton after 8 days when stored at ambient temperature (30°C) but on other hand, the fragrance was retained in β-CD-MCT-treated cotton fabrics for 21 days in the same conditions. Agrawal et.al compared the efficiency of enzymatic treatments and existing chemical techniques for bonding β-CD and its derivatives to cotton surface. Novel chemical based crosslinking with homo-bi-functional reactive dye (C.I. reactive black 5) and grafting with reactive β-CD-MCT show maximum attachment to cotton surface. Innovative, enzymatic coupling of especially synthesized 6-monodeoxy-6-mono(N-tyrosinyl)-β-cyclodextrin was performed on cotton textile surface at low temperature. Alteration in surface topography has been observed for all β-CD treated samples [54]. Martel with co-workers coupled β-CD-MCT to chitosan, to obtain a chitosan derivative bearing cyclodextrin. Because the average degree of substitution of the CD derivative was 2.8, the reaction yielded crosslinked insoluble products. The structure of these materials has been investigated by high-resolution magic-angle spinning (HRMAS) with gradients. For the first time, HRMAS spectra of chitosan polymers containing β-CD were obtained. This NMR technique produced one- and two-dimensional well-resolved solid-state spectra. Decontamination of waters containing textile dyes were carried out with the crosslinked derivatives. Report by Martel showed that the new chitosan derivatives are characterized by a rate of sorption and a global efficiency superior to that of the parent chitosan polymer and of the well-known cyclodextrin-epichlorohydrin gels [55]. El-Tahlawy with co-workers carried out a novel technique for preparation of cyclodextrin-grafted chitosan. β-CD citrate was synthetized by esterifying of β-CD with citric acid (CA) in presence or absence of sodium hypophosphite as a catalyst in a semidry process. β-CD/grafted chitosan was prepared by coupling β-CD citrate with chitosan dissolved in different formic acid solutions having different concentrations. The reacting ingredients were subjected to various reaction conditions to attain the optimum condition. β-CD/grafted chitosan were evaluated by measuring the nitrogen content of both chitosan and grafted chitosan. Chitosan and β-CD/grafted chitosan, having different molecular weights, were evaluated as antimicrobial agents for different microorganisms [56].

Very effective bonding of CDs on cellulose fibres can be achieved by a high-performance resin finish [57] or with non-formaldehyde reagents such as polycarboxylic acids [58, 59] which can covalently esterify hydroxyl groups of cellulose and CDs and therefore the grafting of CD on cellulose can occur. The same linking/crosslinking reagents can be used in the treatment of different synthetic fibres. Polyester fibres were modified by β-CD using citric acid [7, 59] in research work of our group [60], 1,2,3,4-butane tetracarboxylic acid was used as a linker. Odour control is a very important topic in the apparel and underwear items. Odour can be controlled by applying an antimicrobial finish, removing the odour molecules as they are formed or covering up the odour with a fragrance. The odour molecules being hydrophobic become trapped in the cavities of the CDs and are removed during laundering.

Within our research work PET fibres were treated in aqueous solution of different concentrations of β–CD and BTCA. We reported that BTCA molecules react via anhydride formation with hydroxyl groups of β-CD and form nano-assembly which can be physically attached to the PET fibres surface at the elevated temperature. Such assembly could be schematically presented as shown in Figure 3.

For reducing the termofixation temperature the catalyst cyanamide was used. We concluded that the treatment of PET with β-CD/BTCA was very successful even at temperature as low as 115°C when CA as a catalyst has been used. After 10 washings the gain on mass remained as high as 7.8% (Figure 4).

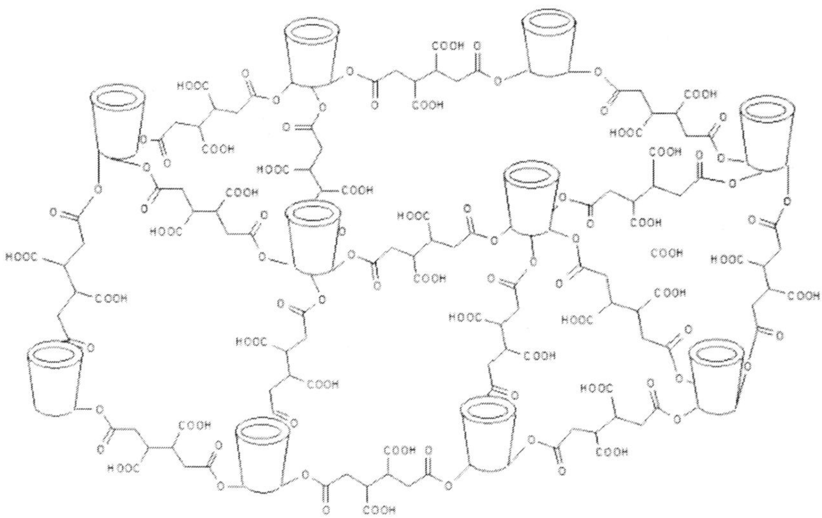

Figure 3. Nano-assembly of β-CD crosslinked with BTCA on textile surface.

We were able to reduce the curing temperature to 115ºC and prepared nanoencapsulated textile materials with increased adsorption capacity (the adsorption of ammonia gas onto treated and untreated PET textile materials was measured using Japan standard test method - JIS K0804) and with postponed release of volatile compounds. From the ammonia gas adsorption measurements (Table 1), it is possible to conclude that the adsorption of ammonium gas increased when PET fabric was treated with β-CD/BTCA/CA at 115ºC/10min - after one hour of exposure to ammonia gas the concentration of gas in the chamber was zero, compare to the concentration when untreated PET fabric was exposed to the ammonium gas, where the concentration in the chamber was changed from the initial value of 125ppm to 77ppm.

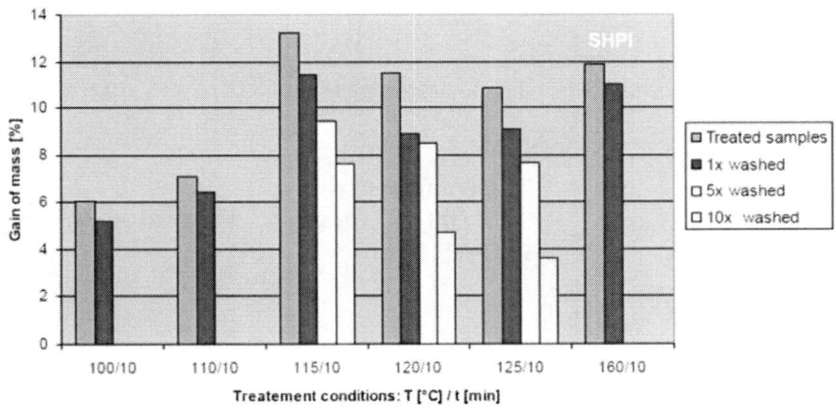

Figure 4. Samples 100/10, 110/10, 115/10, 120/10 and 125/10 are PET samples treated with β-CD, BTCA, CA at 100, 110, 115, 120 and 125°C, respectively; sample 160/10 was treated with β-CD, BTCA and SHPI at 160°C; all samples were termofixed for 10 minutes.

In order to quantify the odour-releasing behaviour of β-CD treated PET fabrics, we organized a sensory panel of nine people to whom the odour was presented under controlled conditions. In order to study the postponed release of the volatile compounds from the β-CD treated textile substrate the following was performed: β-CD/BTCA/CA treated PET textile substrate was sprayed with perfume and dried; the intensity of the perfume from the untreated PET fabric spayed with perfume was also monitored for comparison purposes. Both treatments were performed in triplets. The size of the clothes was 10 by 10 cm. All with perfume treated textile samples were stored separately in dark places. Samples were stored in open conditions so that the perfume was able to evaporate constantly. The odour release was measured once per week. The smell intensity was evaluated from 0 to 4, where 0 means no smell and 4 means very intensive smell. From Figure 5 it is possible to see that the odour release intensity of untreated PET fabrics sprayed with perfume (BLIND, spray) starts to decrease after 6 weeks; the odour intensity of perfume sprayed on β-CD treated fabrics remains constant, but there is slight indication that the intensity of the perfume starts to increase after 6 weeks. We can conclude that, in the case of β-CD treated PET fabrics, some postponed release of the fragrance occurs.

Table 1. Decrease of ammonium gas concentration due to the adsorption.

	PET treated with β-CD/BTCA/ CA at 115ºC/10 min	**Untreated PET**
Initial concentration (ammonia)	125ppm	125ppm
One hour concentration (ammonia)	0ppm	77ppm

Glycidyl methacrylate is widely used in the production of polymer coatings and finishes, adhesives, plastics and elastomers, it can be grafted to various textiles substrates as well. Desmet and co-workers functionalized cotton-cellulose by gamma-irradiation-induced grafting of glycidyl methacrylate (GMA) to obtain a hydrophobic cellulose derivative with epoxy groups suitable for further chemical modification.

Two grafting techniques were applied. In pre-irradiation grafting (PIG) cellulose was irradiated in air and then immersed in a GMA monomer solution, whereas in simultaneous grafting (SG) cellulose was irradiated in an inert atmosphere in the presence of the monomer. In the paper authors claimed that the PIG led to a more homogeneous fibre surface, while SG resulted in higher grafting yield but showed clear indications of some GMA-homopolymerization. Effects of the reaction parameters (grafting method, absorbed dose, monomer concentration, solvent composition) were evaluated by SEM, gravimetry (grafting yield) and FT-IR spectroscopy. It is reported that water uptake of the cellulose decreased while adsorption of pesticide molecules increased upon grafting. The adsorption was further enhanced by β-CD immobilization during SG. This method can be applied to produce adsorbents from cellulose based agricultural wastes [61]. CDs can be incorporated into fibres during the spinning processes [62, 63].

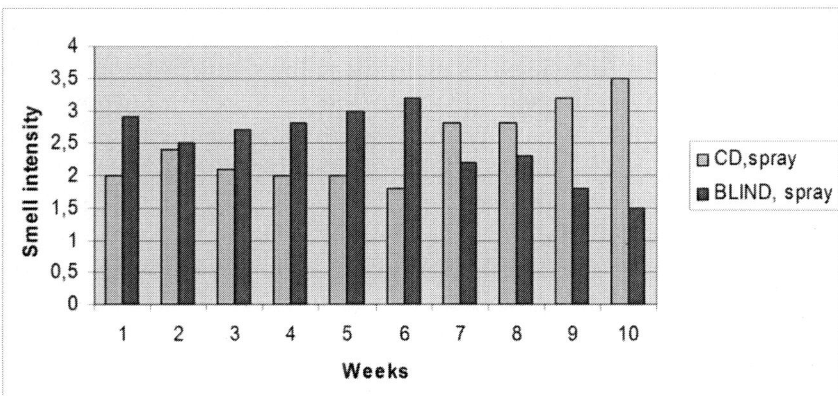

Figure 5. Odour intensity of PET fabrics pre-treated with β-CD (blue) and untreated PET fabrics sprayed with the perfume.

Electrospinning is proven to be an effective method for producing non-woven mats of fibres with high aspect ratios. Manasco and co-workers reported the preparation of submicron hydroxypropyl-beta-cyclodextrin (HP-β-CD) fibres by electrospinning without the addition of a carrier polymer. They focused on exploring solution properties that make fibre formation possible contrary to the widely accepted premise that molecular entanglement of macromolecules is required for electrospinning. The ability to electrospin from these solutions was attributed to hydrogen-bonded aggregation between HP-β-CD molecules at high concentrations [64]. Further it is reported by Uyar and co-workers that poly(methyl methacrylate) (PMMA) nanofibres containing the inclusion complex forming β-CD were successfully produced by means of electrospinning in order to develop functional nanofibrous webs. Electrospinning of uniform PMMA nanofibres containing different loadings of β-CD (10%, 25% and 50% (w/w)) was achieved. The surface sensitive spectroscopic techniques; X-ray photoelectron spectroscopy (XPS) and time-of-flight secondary ion mass spectrometry (ToF-SIMS) showed that some of the β-CD molecules are present on the surface of the PMMA nanofibres, which is essential for the trapping of organic vapours by inclusion complexation. Direct pyrolysis mass spectrometry (DP-MS) studies showed that PMMA nanowebs containing β-CD can entrap organic vapours such as aniline, styrene and toluene from the surroundings due to inclusion complexation with beta-CD that is present on the fibre surface. The study showed that electrospun nanowebs functionalized with CDs may have the potential to be used as molecular filters and/or nanofilters for the treatment of organic vapour waste and air filtration purposes [65]. Polyvinyl alcohol (PVA) nanowebs incorporating vanillin/cyclodextrin inclusion complex (vanillin/CD-IC) were produced via electrospinning technique by Kayaci and Uyar [66]. The vanillin/CD-IC was prepared with three types of CDs; α-CD, β-CD and γ-CD to find out the most favourable CD type for the stabilization of vanillin. PVA/vanillin/CD-IC nanofibres, having fibre diameters similar to 200 nm, were electrospun from aqueous mixture of PVA and vanillin/CD-IC. The results indicated that vanillin with enhanced durability and high temperature stability was achieved for PVA/vanillin/CD-IC nanowebs due to complexation of vanillin with CD. Additionally, they reported that PVA/vanillin/γ-CD-IC nanoweb was more effective for the stabilization and slow release of vanillin suggesting that the strength of interaction between vanillin and the γ-CD cavity is stronger when compared to α-CD and β-CD.

3.1.2. Cosmetotextiles

Although still in its infancy, the market for cosmetotextiles - often referred to as "wearable skincare" - is set to grow rapidly, and the textile industry is optimistic that further technical developments will open up new markets and create growing business opportunities [67]. Cosmetotextile became a fast growing new branch of specialized micro- or nano-encapsulation textile products, with most patents issued after the 1990 [68]. The physical and chemical properties of the guest molecules encapsulated in CDs can change due to complex formation. Thus, for example, the stability of the complexed molecule against light and oxygen increases and the vapour pressure is reduced. The solubility of slightly soluble molecules increases in a CD complex. All these and further advantages of CDs and their complexes can be used for the formulation of cosmetic products [69]. Cosmetotextile allows the administration of active molecules simply and controllably. It can also be used to change the surface properties of a fabric in order to make it self-cleaning, hydrophobic or lipophobic. The article prepared by Ripoll and co-workers reviews the current state of the art concerning functionalization techniques and the methods used to characterize various functionalized fabric. This review also reveals the surprising lack of publications on the functionalization of textile supports [70].

Moist oils (essential oils, herbal oils, oils from flower seeds) have skin care benefits in that they provide an occlusive layer that lubricates the epidermis, together with a moisturizing effect that helps to prevent excess water loss. Essential oils attributed with a range of properties that help to achieve physical and emotional balance. Besides, one additional advantage of molecular encapsulation is the possibility to reload them 4, 58, 71. Cosmetotextile applications can be used in the treatment of chronic venous insufficiency in legs by means of elastic bandages loaded with natural products which possess flebotonic properties. Cravotto and co-workers 72 have developed an efficient synthetic procedure for the preparation of β-CD-grafted viscose by means of the 2-step ultrasound-assisted reaction. The highly grafted fabric bearing bis-urethane bridged β-CD has been characterized by ATR FT-IR and CP-MAS spectra and by an empiric colorimetric method which used phenolphthalein as the CD guest. They have also developed a suitable cosmetic preparation containing natural substances and extracts (aescin, menthol, Centella asiatica and Ginkgo biloba) to recharge the CD-grafted textile. The efficacy of the new cosmetotextile has been corroborated by in vitro studies of diffusion through membranes, cutaneous permeation and accumulation in porcine skin. Aescin was taken as a reference compound and its concentration in the different compartments was monitored by HPLC analysis. They reported that this cost effective cosmetotextile showed excellent application compliance and was easily recharged.

Electrospun functional polystyrene (PS) fibres containing CD-menthol inclusion complexes are new and advance class of cosmetotextiles. Uyar and co-workers developed functional electrospun fibres containing fragrances/flavours with enhanced durability and stability assisted by CD inclusion complexation. As a model fragrance/flavour molecule they used menthol. CD-menthol inclusion complexes were incorporated in electrospun PS fibres by using three types of CDs: α-CD, β-CD and γ-CD. It is reported that due to complexation of menthol with CDs, the stabilization of menthol was achieved up to 350°C whereas the PS fibres without the CD complex could not preserve volatile menthol molecules. This study suggested that the electrospun fibres functionalized with CD are very effective for enhancing the temperature stability of volatile fragrances/flavours and therefore show potentials for the development of functional fibrous materials [73].

Today the subject of well-being is an area which is receiving much interest, with scent being one of the most important aspects of personal care. The definition of the word aromatherapy is the following: therapeutic uses of fragrances which at least mere volatilize to cure and to mitigate or cure diseases, infection and indisposition by means of inhalation alone [74]. The term aromachology was coined in 1982 to denote the science that is dedicated to the study of the relationship between psychology and fragrance technology to elicit a variety of specific feelings and emotions – such as relaxation, exhilaration, sensuality, happiness and well-being-through odours.

CDs linked on cellulose do not affect the cellulose's properties, and CDs keep their ability to form inclusion complexes with other suitable molecules thus, CDs are the first choice in preparing aromatherapy textiles. Lavender is the most used and most versatile of all the essential oils. It is very useful oil, especially when symptoms are due to a nervous problem. The effects of lemon, camomile, rose, cardamom, clove, and jasmine fragrance oils on human have been confirmed by many research works. The sedative effects for the pharmaceutical and emotional effects of essential oils are listed inTables 2 and 3 respectively [75, 76].

3.1.3. Miscellaneous Applications of Cyclodextrins in Textile Industry

An insect repellent is a substance applied to skin, clothing, or other surfaces which discourages insects from landing or climbing on that surface. Synthetic repellents (such as paradichlorobenzene) tend to be more effective than 'natural' repellents, but on the other hand they are usually toxic. Cedar oil is often used as a natural insect repellent or it is used for its aromatic properties, especially in aromatherapy. Essential oil repellents tend to be short-lived in their effectiveness due to their volatile nature. To prevent the essential oils from evaporating form the textile materials we can encapsulate them in β-CD. In our research group [80, 81] with β-CD, nanoencapsulated wool and PET/wool blend fibres were further treated with cedar oil, which is known for being a natural insect repellent. The complex formation of cedar oil with β-CD was determined by ATR FT-IR spectroscopy. Textile material containing β-CD showed, after being treated with cedar oil, a prolonged moths oppression compared to those textile materials treated only with cedar oil. Table 4 presents the damages to wool and larval conditions according to the time of exposure. No visible damage was observed to the naked eye when β-CD/cedar oil treated wool was exposed to a moth's colony for 2 months. In the control (wool samples treated only with cedar oil) no damage was observed for the first few days, but when the cedar oil had evaporated, the wool cloth was not protected anymore. In contrast, when the cedar oil was encapsulated in the β-CD cavity, evaporation was hindered and resistance to insect pests' activities regarding cedar oil remained.

Table 2. Pharmaceutical effect of essential oils [75, 77, 78].

Effects	Essential Oil
Sedation	Mint, Onion, Lemon, Metasequoia
Coalescence	Pine, Clove, Lavender, Onion, Thyme

Effects	Essential Oil
Diuresis	Pine, Lavender Onion. Thyme, Fennel, Lemon, Metasequoia
Facilitating Menses	Pine, Lavender. Mint, Rosemary, Thyme, Basil, Chamomile, Cinnamon, Lemon
Dismissing sputum	Onion, Citrus, Thyme, Chamomile
Allaying a fever	Ginger, Fennel, Chamomile, Lemon
Hypnogenesis	Lavender, Oregano, Basil, Chamomile
Curing Hypertension	Lavender, Fennel, Lemon, Ylangylang
Be good for stomach	Pine, Ginger, Clove, Mint, Onion, Citrus, Rosemary, Thyme, Fennel, Basil, Cinnamon
Diaphoresis	Pine, Lavender, Rosemary, Thyme, Chamomile, Metasequoia
Expelling wind	Ginger, Clove, Onion, Citrus, Rosemary, Fennel, Lemon

Effects	Essential Oil
Losing weigh	Onion, Cinnamon, Lemon
Relieving pain	Vanilla, Lavender. Mint, Onion, Citrus, Rosemary, Chamomile, Cinnamon, Lemon
Detoxification	Lavender
Curing diabetes	Vanilla, Onion, Chamomile, Lemon
Stopping diarrhea	Vanilla, Ginger, Clove, Lavender, Mint, Onion, Oregano, Rosemary, Thyme, Chamomile, Cinnamon, Lemon
Curing flu	Pine, Lavender, Mint, Onion, Citrus, Rosemary, Thyme, Chamomile, Cinnamon, Metasequoia
Curing rheumatism	Lavender, Onion, Citrus, Rosemary, Thyme, Metasequoia
Urging sexual passion	Pine, Ginger, Clove, Mint, Onion, Rosemary, Thyme, Fennel
Relieving spasm	Cinnamon
Promoting	Clove, Lavender, Mint, Onion, Citrus, Rosemary,

Effects	Essential Oil
appetite	Fennel, Basil, Chamomile, Cinnamon, Lemon, Metasequoia
Relieving cough	Rosemary

Table 3. The sedative or emotion effects of essential oils [75, 79].

Emotion	Essential Oils with the Sedative Effects
Anxiety	Benzoin, Lemon, Chamomile, Rose, Cardamom, Clove, Jasmine
Lament	Rose
Stimulation	Camphor, Balm oil
Anger	Chamomile, Balm oil, Rose, Ylangylang
Wretchedness	Basil, Cypress, Mint, Patchouli
Allergy	Chamomile, Jasmine, Balm oil
Distrustfulness	Lavender

Emotion	Essential Oils with the Sedative Effects
Tension	Camphor, Cypress, Vanilla, Jasmine, Balm oil, Lavender, Sandalwood
Melancholy	Basil, Lemon, Chamomile, Vanilla, Jasmine, Lavender, Mint, Rose
Hysteria	Chamomile, Balm oil, Lavender, Jasmine
Mania	Basil, Jasmine, Pine
Irritability	Chamomile, Camphor, Cypress, Lavender
Desolation	Jasmine, Pine, Patchouli, Rosemary

Table 4. The estimation of wool damage and the condition of larval colony in accordance to elapsed time.

Time	β-CD/cedar oil treated wool	Cedar oil treated wool	Untreated wool
48h	No detectable damage	No detectable damage	Very slight visible damage
	Larval	Larval	Larval

Time	β-CD/cedar oil treated wool	Cedar oil treated wool	Untreated wool
	conditions: live	conditions: live	conditions: live
72h	No detectable damage	No detectable damage	Moderate visible damage
	Larval conditions: dead	Larval conditions: dead	Larval conditions: live
7 days	Addition of new larvae	Addition of new larvae	/
14 days	No detectable damage	Very slight visible damage	Very heavy damages
	Larval conditions: dead	Larval conditions: live	Larval conditions: live
56 days	No detectable damage	Moderate visible damages	Very heavy damages
	Larval conditions: dead	Larval conditions: live	Larval conditions: live, pupating

To maintain antimicrobial activity, frequent administration of conventional formulations of many antibiotics with short half-life is necessary. To enhance release properties, many materials have been introduced into the matrix and coating extended-release system in the past few years. The review by Gao [82] highlights the development of materials used in extended-release formulation and nanoparticles for antibiotic delivery. CDs are mentioned as nanoparticles/nanocarriers which allow the antibiotic extended-release.

A textile polyester vascular graft can be modified by methyl-β-CD to obtain a new implant capable of releasing antibiotics directly in situ at the site of operation over a prolonged period and thereby preventing post-operative infections [83]. Wang reported the inclusion complex of miconazole nitrate with β-CD formation by the co-precipitation method. The DSC curve and X-ray diffraction verified the inclusion complex formation between β-CD and miconazole nitrate. The skin-care textiles can be obtained by treating fabrics with inclusion complexes using the sol-gel method [84].

Novel nano-porous polymers or nanosponges can be prepared for removal of organic pollutants from waste water. The polymeric «nanosponge» materials are not durable (usually they are in gel form), they do not have high mechanical strength, so they must be impregnated onto the pore structure of a ceramic or some other porous surfaces [85, 86]. This technology is very specific for the target pollutant, it is very expensive and the removal of the adsorbed pollutant from the nanosponge is not possible. Textile materials are very important as filter materials. The cost of textile materials is acceptable (polyester, viscose), they have sufficient mechanical strength; the pore size, especially the macro-pore size can vary and it depends on the type of textile (the density of non-woven material) and on the diameter of the fibres. Textile materials can be further modified to prepare filtration materials with additional adsorption.

The amount of aromatic organic pollutants (phenols, aniline, formaldehyde and others) can be reduced from dyeing wastewater by using CDs which can be immobilized on a water insoluble organic support. The new concept for modification of textile substrates based on permanent fixation of supramolecular compounds - CDs on the material surface thus imparts new functionality to the fabric [87]. The guest molecules could be various organic molecules and some metal ions as well. The assembly of nanocapsules on textile materials acts as selective filtration/adsorption media for various pollutants. Prabaharan and Mano further reported that CDs have recently been recognized as useful adsorbent matrices. Due to its hydrophobic cavity, CDs can interact with appropriately sized molecules to result in the formation of inclusion complexes. These complexes are of interest for scientific research as they exist in aqueous solution and can be used to study the hydrophobic interactions which are important in the biomedical and environmental fields. The grafting of CD onto chitosan can result in the formation of a molecular carrier that possess the cumulative effects of inclusion, size specificity and transport properties of CDs as well as the controlled release ability of the polymeric matrix. In this review, different methods of CD grafting onto chitosan are discussed [88]. Electrospinning has been used to create polystyrene (PS) nanofibres containing any of the three different types of cyclodextrin (CD); α-CD, β-CD, and γ-CD [89]. These three CDs are chosen because they have different sized cavities that potentially allow for selective inclusion complex (IC) formation with molecules of different sizes or differences in affinity of IC formation with one type of molecule. The comparative efficiency of the PS/CD nanofibres/nanoweb for removing phenolphthalein, a model organic compound, from solution was determined by UV-Vis spectrometry, and the kinetics of phenolphthalein capture was shown to follow the trend PS/α-CD > PS/β-CD > PS/γ-CD. Direct pyrolysis mass spectrometry (DP-MS) was also performed to ascertain the relative binding strengths of the phenolphthalein for the CD cavities, and the results showed the trend in the interaction strength was β-CD > γ-CD > α-CD. Results of their research demonstrated that nanofibres produced by electrospinning that incorporate CDs with different sized cavities can indeed filter organic molecules and can potentially be used for filtration, purification, and/or separation processes.

4. CONCLUSION

Since 1980, when Szejtli first patented the bonding of CDs onto textile fibres, a lot of research has gone into the application of CDs on to textile substrates. But there is still a gap between original high level basic science and commercial applications of CDs in all industrial sectors.

Nevertheless the use of CDs in the textile industry has increased in the last years. Grafted CDs on textile substrates or spun fibres which contain

CDs can be used to obtain special functionality of textiles such as absorption; they can complex and release fragrances or "skin-care-active" substances like vitamins, caffeine and menthol as well as bioactive substances such as biocides and insecticides and drugs. Furthermore, various textile materials treated with CDs could be used for adsorption of small pollutants from waste waters - for filtration, purification, and/or separation treatments of waste waters.

REFERENCES

1. Vögtle F. Supramolecular Chemistry, an introduction. New York: John Wiley & Sons; 1991.

2. Astray G, Gonzalez-Barreiro C, Mejuto JC, Rial-Otero R, Simal-Gándara J. A review on the use of cyclodextrins in foods. Food Hydrocolloids 2009;23(7) 1631–1640.

3. Jeang CL, Lin DG, Hsieh SH. Characterization of cyclodextrin glycosyltransferase of the same gene expressed from Bacillus macerans, Bacillus subtilis, and Escherichia coli. Journal of Agricultural and Food Chemistry 2005;53(16) 6301-6304.

4. Szejtli J. Cyclodextrins in the Textile Industry. Starch/Stärke 2003;55(5) 191-196.

5. Cravotto G, Binello A, Baranelli E, Carraro P, Trotta F. Cyclodextrins as food additives and in food processing. Current Nutrition & Food Science 2006;2(4) 343–350.

6. Dastjerdi R, Montazer M. A review on the application of inorganic nano-structured materials in the modification of textiles: Focus on anti-microbial properties. Colloids and Surfaces B: Biointerfaces 2010;79(1) 5-18.

7. Martin del Valle EM. Cyclodextrins and their uses: a review. Process Biochemistry 2004;39(9) 1033–1046.

8. Li S, Purdy WC. Cyclodextrins and their Applications in Analytical-Chemistry. Chemical Reviews 1992;92(6) 1457-1470.

9. Szejtli J. Introduction and General Overview of Cyclodextrin Chemistry. Chemical Reviews 1998;98(5) 1743–1753.

10. Uekama K, Hirayama F, Irie T. Cyclodextrin Drug Carrier Systems. Chemical Reviews 1998;98(5) 2045-2076.

11. Ribeiro ACF, Valente AJM, Lobo VMM. Transport Properties of Cyclodextrins: Intermolecular Diffusion Coefficients. Journal of the Balkan Tribological Association 2008;14(3) 396-404.

12. Harada Laboratory. Department of Macromolecular Science. Osaka University. http://www.chem.sci.osaka-u.ac.jp/lab/harada/eng/eng/research/01.html (accessed 2 July 2012)

13. Saenger W. Sterochemistry of circularly closed oligosaccharides: cyclodextrins structure and function. Biochemical Society Transactions 1983;11(2) 136-139 Connors, K. A. (1997). The Stability of Cyclodextrin Complexes in Solution. Chemical Reviews, Vol.97, No.5, 325-1357, ISSN: 0009-2665

14. Connors KA. The Stability of Cyclodextrin Complexes in Solution. Chemical Reviews 1997;97(5) 325-1357.

15. Weber E. Molecular Inclusion and Molecular Recognition - Clathrates I. Topics in Current Chemistry 1987;140 1-20.

16. Montazer M, Jolaei MM. β-Cyclodextrin stabilized on three-dimensional polyester fabric with different crosslinking agents. Journal of Applied Polymer Science 2010;116(1) 210–217.

17. Lo Meo P, D'Anna F, Riela S, Gruttadauria M, Noto R. Spectrophotometric study on the thermodynamics of binding of α- and β-cyclodextrin towards some p-nitrobenzene derivatives. Organic & Biomolecular Chemistry 2003;1(9) 1584-1590.

18. Voncina B. Application of cyclodextrins in textile dyeing. In: Hauser PJ (ed.) Textile dyeing. Rijeka: InTech; 2011. p373-392.

19. Fernandes CM, Carvalho RA, da Costa SP, Veiga FJB. Multimodal molecular encapsulation of nicardipine hydrochloride by β-cyclodextrin, hydroxypropyl-β-cyclodextrin and triacetyl-β-cyclodextrin in solution. Structural studies by 1H NMR and ROESY experiments. European Journal of Pharmaceutical Sciences 2003;18(5) 285-296.

20. Figueiras A, Sarraguca JMG, Carvalho RA, Pais AACC, Veiga FJB. Interaction of Omeprazole with a Methylated Derivative of β-Cyclodextrin: Phase Solubility, NMR Spectroscopy and Molecular Simulation. Pharmaceutical Research 2007;24(2) 377-389.

21. Monti S, Sortino S. Photoprocesses of photosensitizing drugs within cyclodextrin cavities. Chemical Society Reviews 2002;31(5) 287-300.

22. Nilsson M, Cabaleiro-Lago C, Valente AJM, Soderman O. Interactions between gemini surfactants, 12-s-12, and beta-cyclodextrin as investigated by NMR diffusometry and electric conductometry. Langmuir 2006;22(21) 8663-8669.

23. Cabaleiro-Lago C, Nilsson M, Soderman O. Self-diffusion NMR studies of the host-guest interaction between beta-cyclodextrin and alkyltrimethylammonium bromide surfactants. Langmuir 2005;21(25) 11637-11644.

24. Valente AJM, Nilsson M, Soderman O. Interactions between n-octyl and n-nonyl β-d-glucosides and α- and β-cyclodextrins as seen by self-diffusion NMR. Journal of Colloid and Interface Science 2005;281(1) 218-224.

25. Garcia-Rio L, Godoy A. Use of spectra resolution methodology to investigate surfactant/β-cyclodextrin mixed systems. Journal of Physical Chemistry B 2007;111(23) 6400-6409.

26. Voncina B, Vivod V, Jausovec D. [Beta]-cyclodextrin as a retarding reagent in polyacrylonitrile dyeing. Dyes and pigments 2007;74(3) 642-646.

27. Costa T, Seixas de Melo JS. The effect of γ-cyclodextrin addition in the self-assembly behavior of pyrene labeled poly(acrylic) acid with different chain sizes. Journal of Polymer Science Part a-Polymer Chemistry 2008;46(4) 1402-1415.

28. Denter U, Schollmeyer E. Surface modification of synthetic and natural fibres by fixation of cyclodextrin derivatives. Journal of Inclusion Phenomena and Molecular Recognition in Chemistry 1996;25(1-3) 197-202.

29. Szejtli J. Utilization of cyclodextrins in industrial products and processes. Journal of Materials Chemistry 1997;7(4) 575–587.

30. Frömming KH, Szejtli, J. Cyclodextrins in Pharmacy. Dordrecht: Kluwer Academic Publishers; 1994.

31. Hedges AR. Industrial applications of cyclodextrins. Chemical Reviews 1998;98(5) 2035-2044.

32. Buschmann HJ, Knittel D, Schollmeyer E. Textile Materialien Mit Cyclodextrinen, Textile Materials with Cyclodextrins. Melliand Textilberichte 2001;82(5) 368-370.

33. Buschmann HJ. Cosmetic textiles: Clothes with a skin care action. Cossma 2001;2(7) 38-39.

34. Buschmann HJ, Knittel D, Beermann K, Schollmeyer E. Cyclodextrins and textiles. Nachr. Chem. 2001;49(5) 620-620.

35. Murthy CN, Shown I. Grafting of Cotton Fiber by Water-Soluble Cyclodextrin-Based Polymer. Journal of Applied Polymer Science 2009;111(4) 2056-2061.

36. Savarino P, Viscardi G, Quagliotto P, Montoneri E, Barni E. Reactivity and effects of cyclodextrins in textile dyeing. Dyes and Pigments, 1999;42(2) 143-147,.

37. Poulakis, K, Buschmann HJ, Schollmeyer E. Ger. Offen. 4035378. 1992; WO 2002046520. 2002.

38. Buschmann HJ, Knittel D, Schollmeyer E. Resin finishing of cotton in the presence of cyclodextrins for depositing fragrances. Melliand Textilberichte 1991;72(3) 198-199.

39. Fujimura T. Jpn. Kokai Tokkyo Koho 60259648. 1985.

40. Ritter W, Delney J, Volz W, Kerr I. DE 10101294. 2002.

41. Akasaka M, Shibata T, Ochia H. Jpn. Kokai Tokkyo Koho 03059178. 1991.

42. Akasaka, M, Sawai Y, Iwase K, Moriishi H. EP 488294. 1992.

43. Yamamoto K, Saeki T. Jpn. Kokai Tokkyo Koho 09228144. 1997.

44. Yamamoto K, Saeki T. Jpn. Kokai Tokkyo Koho 10007591. 1998.

45. Voncina B, Majcen N. Application of cyclodextrin for medical and hygienic textiles. Tekstil 2004;53(1) 1-9.

46. Knittel D, Schollmeyer E. Technologies for a new century. Surface modification of fibres. Journal of the Textile Institute 2000;91(3) 151-165.

47. Szejtli J, Zsadon B, Fenyvesi E, Otta K, Tudos F. Hungarian Patent: 181733. 1980. US Patent 4,357,468 (1982).

48. Szejtli J, Zsadon B, Horvath OK, Ujhazy A, Fenyvesi E. Hungarian Patent: 54506. 1991.

49. Reuscher H, Hirsenkorn R, Haas W. German Patent: DE 19520967. 1996.

50. Grechin AG, Buschmann HJ, Schollmeyer E. Quantification of Cyclodextrins fixed onto Cellulose Fibres. Textile Research Journal 2007;77(3) 161-164.

51. Moldenhauer JP, Reuscher H. Textile finishing with MCT-beta-cyclodextrin. In: Torres Labandeira JJ, Vila-Jato JL. (eds.) Proceedings of the 9th International Symposium on Cyclodextrins, 31 May – 3 June 1998, Santiago de Compostela, Spain. Dordrecht: Kluwer Academic Publishers; 1999.

52. Ibrahim NA, Allam EA, El-Hossamy MB, El-Zairy WM. UV-Protective Finishing of Cellulose/Wool Blended Fabrics. Polymer-Plastics Technology and Engineering 2007;46(9), 905-911.

53. Sricharussin W, Sopajaree C, Maneerung T, Sangsuriya N. Modification of cotton fabrics with beta-cyclodextrin derivative for aroma finishing. Journal of the Textile Institute 2009;100(8) 682-687.

54. Agrawal PB, Warmoeskerken MMCG. Permanent fixation of β-cyclodextrin on cotton surface: An assessment between innovative and established approaches. Journal of Applied Polymer Science 2012;124(5) 4090-4097.

55. Martel B, Devassine M, Crini G, Weltrowski M, Bourdonneau M, Morcellet M. Preparation and sorption properties of a beta-cyclodextrin-linked chitosan derivative. Journal of Polymer Science Part A-Polymer Chemistry 2001;39(1) 169-176.

56. El-Tahlawy K, Gaffar MA, . Novel method for preparation of beta-El-Rafie Scyclodextrin/grafted chitosan and it's application. Carbohydrate Polymers 2006;63(3) 385-392.

57. Ostertag H. Anwendung von β-Cyclodextrinen in der CO-Gewebeveredlung. Melliand Textilberichte 2002;83(11-12) 872-878.

58. Voncina B, Le Marechal AM. Grafting of cotton with beta-cyclodextrin via poly(carboxylic acid). Journal Of Applied Polymer Science 2005;96(4) 1323-1328.

59. Martel B, Weltrowski M, Ruffin D, Morcellet M. Polycarboxylic acids as crosslinking agents for grafting cyclodextrins onto cotton or wool fabrics: Study of the process parameters. Journal of Applied Polymer Science 2002;83(7) 1449-1456.

60. Vončina B, Le Marechal AM. [Beta]-cyclodextrin in medical and hygienic textiles. In: Chen X, Ge Y, Yan X. (eds.). 83rd TIWC. Quality textiles for quality life: proceedings of the Textile Institute 83rd World Conference (83rd TIWC), 23-27 May 2004, Shanghai, China. Manchester: Shanghai: Textile Institute, Donghua University; 2004.

61. Desmet G, Takacs E, Wojnarovits L, Borsa J. Cellulose functionalization via high-energy irradiation-initiated grafting of glycidyl methacrylate and cyclodextrin immobilization. Radiation Physics and Chemistry 2011;80(12) 1358-1362.

62. He Y, Inoue Y. Effect of alpha-cyclodextrin on the crystallization of poly(3-hydroxybutyrate). Journal of Polymer Science Part B-Polymer Physics 2004;42(18) 3461-3469.

63. Vogel R, Tandler B, Haussler L, Jehnichen D, Brunig H. Melt spinning of poly(3-hydroxybutyrate) fibers for tissue engineering using alpha-cyclodextrin/polymer inclusion complexes as the nucleation agent. Macromolecular Bioscience 2006;6(9) 730-736.

64. Manasco JL, Saquing CD,Tang C,Khan SACyclodextrin fibers via polymer-free electrospinning. RSC Advances 2012;2(9) 3778-3784

65. Uyar T, Havelund R,Nur Y,Balan A , Hacaloglu JToppare L, Besenbacher F,Kingshott P. functionalized poly(methyl methacrylate) (PMMA) electrospun nanofibers for organic vapors

waste treatment. Journal of Membrane Science 2010;365(1-2) 409-417.

66. Kayaci F, Uyar T. Encapsulation of vanillin/cyclodextrin inclusion complex in electrospun polyvinyl alcohol (PVA) nanowebs: Prolonged shelf-life and high temperature stability of vanillin Food Chemistry 2012;133(3) 641-649

67. Hague L. Cosmetotextiles market has exciting potential. Aroq Ltd http://www.just-style.com/analysis/cosmetotextiles-market-has-exciting-potential_id111926.aspx (accessed 3 July 2012).

68. CEN standard CEN/TC 248 Textiles and textile products; WG 25 Cosmeto-textiles. CEN - ISO standardization committees on textiles Cheng S.Y. "Development of Cosmetic Texiles Using Microencapsulation Technology", RJTA, vol.12, no.4, 2008

69. Buschmann HJ, Schollmeyer E. Applications of cyclodextrins in cosmetic products. Journal of Cosmetic Science 2002;53(3) 185-191.

70. Ripoll L, Bordes C, Etheve S, Elaissari A, Fessi H. Cosmeto-textile from formulation to characterization: an overview. e-Polymers 2010;040 http://www.e-polymers.org/journal/papers/ hfessi_010410.pdf (accessed 5 July 2012).

71. Buschmann HJ, Knittel D, Schollmeyer E. New Textile Applications of Cyclodextrins. Journal of Inclusion Phenomena and Macrocyclic Chemistry 2001;40(3) 169-172.

72. Cravotto G, Beltramo L, Sapino S, Binello A, Carlotti ME. A new cyclodextrin-grafted viscose loaded with aescin formulations for a cosmeto-textile approach to chronic venous insufficiency. Journal of Materials Science-Materials in Medicine 2011;22(10) 2387-2395

73. Uyar T, ,Hacaloglu J .Besenbacher F Electrospun polystyrene fibers containing high temperature stable volatile fragrance/flavor facilitated by cyclodextrin inclusion complexes. Reactive & Functional Polymers 2009;69(3) 145-150.

74. Buchbauer G. Aromatherapy: Use of fragrance and essential oils as medicaments. Flavour and Fragrance Journal 1994;9 217-222.

75. Wang CX, Chen SH. Aromachology and Its Application in the Textile Field. Fibres & Textiles in Eastern Europe 2005;13(6(54)) 41-44.

76. Buschmann HJ. Removal of Residual Surfactant Deposits from Textile Materials with the Aid of Cyclodextrins. Melliand Textilberchite. 1995(9).

77. Jellinnek JS. Aromachology: A Status Review. Perfume & Flavorist 1994;19 25-49.

78. Wu CH. Essential Oil and Aromatherapy. Kexue Nong Ye 1999;47(3) 1-3.

79. Mazzaro D. The home fragrance market. Chemical Market Reporter 2000;4 14.

80. Vraz Kresevic S, Voncina B, Vukusic Bischof S, Katovic D. Textile Materials Treated With Eco - Friendly Insecticide Agents. In: Dragcevic Zvonko (ed.): 4th International Textile, Clothing & Design Conference: Magic world of textiles: book of proceedings, ITC&DC, 05-08 October, 2008, Dubrovnik, Croatia. Zagreb: Faculty of Textile Technology, University of Zagreb; 2008.

81. Vraz Kresevic S, Voncina B, Gersak J. Insect resistant and eco-friendly textile materials. In: Dragcevic Zvonko (ed.): 4th International Textile, Clothing & Design Conference: Magic world of textiles : book of proceedings, ITC&DC, 05-08 October, 2008, Dubrovnik, Croatia. Zagreb: Faculty of Textile Technology, University of Zagreb; 2008.

82. Gao P, Nie X, Zou MJ, Shi YJ, Cheng G. Recent advances in materials for extended-release antibiotic delivery system. Journal of Antibiotics 2011;64(9) 625-634.

83. Blanchemain N, Karrout Y, Tabary N, Neut C, Bria M, Siepmann J, Hildebrand HF, Martel B. 2011Methyl-beta-cyclodextrin modified vascular prosthesis: Influence of the modification level on the drug delivery properties in different media. Acta Biomaterialia 2011;7(1) 304-314.

84. Wang JH, Cai ZS. A study of inclusion complex of miconazole nitrate with beta-cyclodextrin and its application on protein fabrics. In: Bai L, Rui Y. (eds.) Researches and Progresses of Modern Technology on Silk, Textile And Mechanicals I: Conference publication of the 6th

China International Silk Conference and the 2nd International Textile Forum, 13-14 September 2007, Suzhou, Jiangsu, P.R. China. Chemical Industry Press; 2007.

85. Salipira KL, Krause RW, Mamba BB, Malefetse TJ, Cele LM, Durbach SH. Cyclodextrin polyurethanes polymerised with multi-walled carbon nanotubes: Synthesis and characterisation. Materials Chemistry and Physics 2008;111(2-3) 218-224.

86. Salipira KL, Mamba BB, Krause RW, Malefetse TJ, Durbach SH. Cyclodextrin polyurethanes polymerised with carbon nanotubes for the removal of organic pollutants in water. Water SA 2008;34 113-118. http://www.wrc.org.za/downloads/watersa/2008/2198.pdf (accessed 3 July 2012).

87. Mamba BB, Krause RW, Malefetse TJ, Mhlanga SD, Sithole SP, Salipira KL, Nxumalo EN. Removal of geosmin and 2-methylisorboneol (2-MIB) in water from Zuikerbosch Water Treatment Plant (Rand Water) using β-cyclodextrin polyurethanes. Water SA 2007;33 223-228. http://www.wrc.org.za/downloads/watersa/2007/Apr%2007/2002.pdf (accessed 3 July 2012).

88. Prabaharan M, Mano JF. Chitosan derivatives bearing cyclodextrin cavities as novel adsorbent matrices. Carbohydrate Polymers 2006;63(2) 153-166.

89. Uyar T, Havelund R, Hacaloglu J, Besenbacher F, Kingshott P. Functional Electrospun Polystyrene Nanofibers Incorporating alpha-, beta-, and gamma-Cyclodextrins: Comparison of Molecular Filter Performance. ACS Nano 2010;4(9) 5121-5130.

CHAPTER 7

Extraction of eco-friendly natural dyes from mango leaves and their application on silk fabric

Mohammad Gias Uddin

Department of Textile Engineering, Faculty of Engineering, Ahsanullah University of Science and Technology, Dhaka 1208, Bangladesh

ABSTRACT

The aim of the study was to evaluate the performance of dyes extracted from mango leaves in silk dyeing. Extraction medium was optimized by extracting dyes from fixed quantity of crushed leaves under pH values from 3 to 12. The maximum relative color strength of the extracted dye liquor was found to be at pH 10. The optimum dye extraction conditions i.e., the temperature, time, and material-to-liquor ratio were found to be 98 °C, 60 min, and 1:10, respectively. Dyeing was carried out with the optimized dye extract on mordanted and unmordanted silk fabrics. The dyed materials were evaluated by measuring the color yield and fastness properties. It was concluded that the color values were found to be influenced by the addition of mordants, consequently different fashion hues were obtained from the same dye extract using different mordants. It can also be said that mango leaves have good potentiality for dyeing of silk fabric.

1. BACKGROUND

Textile dyeing industry at present uses excessive amount of synthetic dyes to meet the required coloration of global consumption of textiles due to cheaper prices, wider ranges of bright shades, and considerably improved fastness properties in comparison to natural dyes (El-Nagar et al. 2005; Iqbal et al. 2008). But the production of synthetic dyes is dependent on petrochemical source, and some of these dyes contain carcinogenic amines (Hunger 2003). The application of such dyes causes serious health hazards and influences negatively the eco-balance of nature (Bruna and Maria 2013; Goodarzian and Ekrami 2010; Jothi 2008). Moreover, many countries already imposed stringent environment standards over these dyes. For instance, Germany has banned the azo dyes (Almahy et al. 2013). In this situation, a higher demand is put towards the greener alternatives or agricultural residues (Ammayappan et al. 2014). As a result, natural dyes are among the promising options for developing a greener textile dyeing process and such interest is reflected to the increased number of recent publications. Plant leaves are potential sources of natural dyes because of their easy availability and abundant nature.

Silk has been known as the "queen of fibers" since its discovery. Clothes made from silk are luxurious and have many excellent qualities including the material's luster, light weight, superior mechanical performance, fine and smooth texture, excellent moisture transportation, and excellent draping quality (Cai et al. 2001). Mango bark has been reported to be used on silk and cotton materials as a source of natural dyes, and a wide range of colors have been produced using different mordants (Bains et al. 2003; Win and Swe 2008). On the other hand, the use of acid activated mango leaf powder (MLP) has been reported in another study for the removal of the Rhodamine B (RB) dye from aqueous solution (Khan et al. 2011). However, apart from this application of mango leaves, different leaves such as peach, poinsettia, acalypha, and parthenium leaves have also been reported to extract colors which were used in dyeing of silk materials (Mahajan et al. 2005;

Rawat et al.2006; Saravanan et al. 2013; Suneeta and Mahale 2002) while mango leaves have been reported to be used in batik painting technique on silk fabric in comparison with other four natural dyes (Klaichoi and Padungtos 2010). There is scope to extract color from mango leaves for the use in dyeing of silk fabric in order to get different fashion hues. The aim of the research was to evaluate the performance of dyes extracted from mango leaves in silk dyeing. The specific objectives were to analyze the aqueous extraction process of the dyes, to explore the possibilities of producing fashionable hues from the dyes using different mordants, to compare between unmordanted and mordanted dyed fabrics, to analyze the color values, and to assess the color fastness properties of dyed fabric.

2. METHODS

2.1 Materials

Mango leaves used for the extraction purpose was collected from Roads & Highways Department, Dhaka, Bangladesh. Mangiferin as shown in Fig. 1 (1,3,6,7-tetrahydroxyxanthone-C-2-β-D-glucoside) was the chemical responsible (Luo et al. 2012) for providing color from mango leaves.

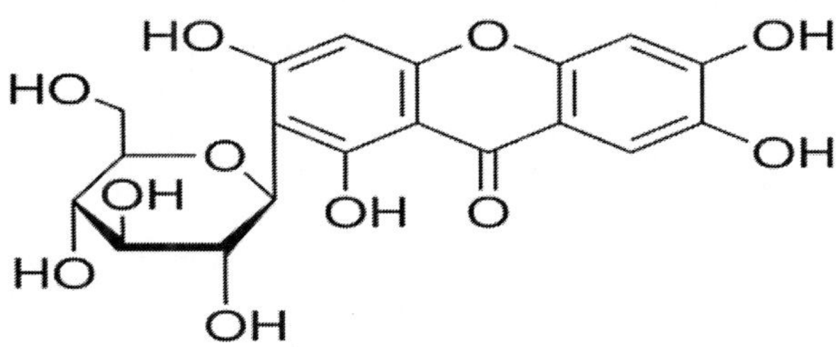

Figure. 1 Chemical structure of mangiferin

Plain weave (1/1) raw silk fabric (22 g/m^2 fabric) purchased from Sopura Silk Limited, Dhaka, was used for this study.

2.1.1 Extraction

The leaves were washed thoroughly with water to remove dirt. They were dried under direct sunlight and grinded into very small units with the help of a grinding machine. The wastages are removed using a fine strainer, and finally, weight was taken. After drying, crushing, and removing wastages, the weight of 1 kilogram leaves was found to be 318 gram. Raw, dried, and crushed leaves are shown in Fig. 2.

Figure. 2 (**A**) Raw, (**B**) dried, and (**C**) crushed mango leaves

The color component was extracted from the leaves in aqueous extraction process. Extraction was carried out with fixed quantity of crushed leaves (10 gram) under ten different pH values from pH 3 to 12 with a liquor ratio of 1:10 (Weight of crushed leaves in gram; amount of water in milliliter) at 98 °C for 60 min to optimize extraction medium. In each process of extraction, the mixture was cooled down and finally the dye extracts were filtered with fine filter paper three times to ensure clear dye solution.

The dye extracts obtained at different pH values were used for obtaining standard calibration curves through their absorbance values found using a dual beam reflectance spectrophotometer. The dilution of the extracts was carried out for the linear dependence on the concentration-absorbance relation at an absorbance peak (λ_{max}). The absorbance values of extracted dye liquors under alkaline (pH 8–12) and acidic (pH 3–6) conditions were considered as batches, and relative color strength values of these batch solutions were measured from the spectrophotometer by comparing with the absorbance value of extracted dye liquor under neutral condition (pH 7) which was considered as standard.

Again, the dye extract which gave the maximum color strength was utilized to optimize the extraction levels of temperature, time, and material-to-liquor ratio. An orthogonal design of experiments was undertaken for this purpose.

2.1.2 Degumming

Raw silk fabric was degummed in an aqueous solution containing soap (15 g/L), sequestering agent (1 g/L), and wetting agent (1 g/L) maintaining the bath at pH 9. The material-to-liquor ratio during the treatment was maintained at 1:50. The temperature was gradually raised to 80 °C and run for 60 min. The degummed fabric was washed with 2 g/L detergent at 65 °C for 10 min.

2.1.3 Bleaching

The degummed fabric was bleached by treating with 35 % hydrogen peroxide (3 mL/L), sequestering agent (1 g/L), wetting agent (1 g/L), and trisodium phosphate (2 g/L), maintaining a material-to-liquor ratio of 1:50 at pH 9 and temperature 60 °C for 60 min followed by washing with 2 g/L detergent at 65 °C for 10 min. CIE whiteness index of the bleached fabric was found to be 63.26.

2.1.4 Mordanting

Pre-mordanting was carried out on silk fabric using 5 % (on fabric weight) of ferrous sulfate, alum (potassium aluminum sulfate), and tin (stannous chloride) mordants individually and using four different combinations of mordants such as ferrous sulfate-alum (2.5 + 2.5 %), ferrous sulfate-alum-tin (2 + 2 + 1 %), alum-tin (2.5 + 2.5 %), and alum-tin-tannic acid (2 + 2 + 1 %) at 60 °C for 60 min keeping a material-to-liquor ratio of 1:30. Again, cream of tartar (CT) was used as a mordant assistant (Mortazavi et al. 2012) with stannous chloride, written as tin-CT.

2.1.5 Dyeing

Dyeing was carried out IR sample dyeing machine with the optimized dye extract as per standard parameters recommended for silk fabric, reported in Clariant manual, i.e., at 80 °C for 60 min under pH 5, keeping a material-to-liquor ratio of 1:50. Opticid PSD (1.5 g/L) was used as a buffering agent in the extracted dye liquor.

2.1.6 Color yield of dyed fabrics

Dyed samples were analyzed by measuring the reflectance curve between 350 and 750 nm with the spectrophotometer with illuminant D_{65} at 10^0 observer. The minimum of the curve (R_{min}) was used to determine the ratio of light absorption (*K*) and scatter (*S*) via the Kubelka-Munk function (Mcdonald 1997).

$$\left(\frac{K}{S}\right)_{\text{Dyed}} = \frac{(1-R_{\min})^2}{2R_{\min}}. \tag{1}$$

2.1.7 Color coordinates of dyed fabrics

The color coordinates of the dyed samples were determined based on the CIELab system via the spectrophotometer. In addition, ΔE_{CMC}value was determined to show the color difference between mordanted and unmordanted samples.

2.1.8 Color fastness

Washing and light fastness tests were carried out in ISO 105 C02 and ISO 105 B02 method, respectively.

3. RESULTS AND DISCUSSION

3.1 Color strength of extracted dye liquors

Optimum pH was selected based on the relative color strength value of the extracted dye liquor at which maximum color was extracted. Changes in color strength were found due to changes in pH as shown in Table 1.

Table 1 Relative color strength of extracted dye liquors at pH 3 to12

pH of extraction bath	Color strength values
3	95
4	88.6
5	88.9
6	95.7
7	100
8	103.3
9	105
10	108.5
11	102.1
12	93.3

It can be seen from the extraction results that the extracted solution showed maximum color strength at pH 10 which was 108.5. It was also found that from the neutral condition (pH 7), relative color strength values gradually decreased up to pH 4 and then increased at pH 3. The reason of extracting more coloring component in alkaline medium was due to the presence of acidic phenolic groups in mangiferin which reacted with alkali and formed more soluble salts in water as shown in Fig. 3. The solubility of the coloring component was increased due to the increased ionization of hydroxyl (phenoxide) groups in alkaline medium (Ali 2007).

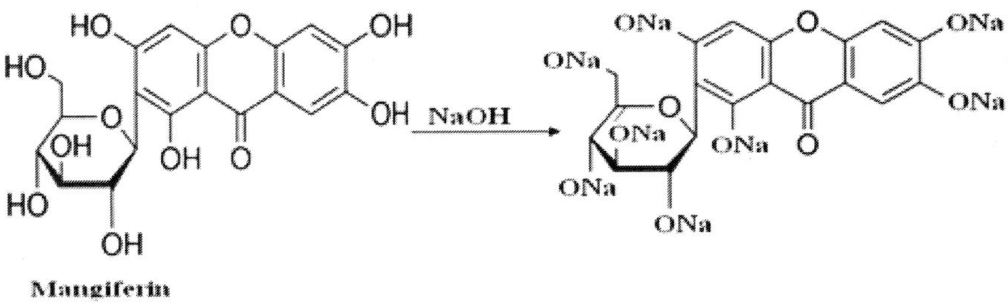

Figure. 3 Reaction of mangiferin with caustic soda

Again, increasing the pH from neutral condition improved the color strength of the extracted dye liquors up to a certain point. A further increase in alkaline pH resulted in decrease in the color strength of the extract. This decline in color strength was due to the high reactivity of mangiferin in concentrated alkaline medium (Spyroudis 2000).

Furthermore, the cell wall of leaves is composed of cellulosic material which gains anionic charge under alkaline medium. Because of these anionic repulsive forces among the cell walls, they lose their strength and ruptured easily (Ali 2007). In addition, as the observed leaf dyes have polyphenolic chromophoric structure, hence better extractions were observed using aqueous method (Sivakumar et al. 2009a, b).

3.2 Study of pH stability of dye extracts

It was noticeable during extraction that pH of the extraction bath changed gradually with time. Table 2 shows the pH variation after filtration and with time elapsed.

Table 2 Changes in pH at the extraction process

Sl. No.	Set pH for extraction	pH after extraction	
		After filtration	After 24 h of filtration
1	3	2.9	3.3
2	4	3.9	4.2
3	5	4.6	4.9
4	6	4.9	5.1
5	7	5.2	5.1
6	8	5.9	5.6
7	9	6.2	5.8
8	10	6.4	6
9	11	6.7	6.7
10	12	7.3	7

pH was found to be decreased in all the extraction baths from pH 3 to 12. This was due to the release of acidic color components from the leaves during extraction. From the neutral pH bath where the pH was set 7, the higher the alkalinity of the extraction bath, the greater was the pH drop rate. The drop rate became gradually slower while gradually approaching to more acidic bath from neutral bath. The pH was also measured after 24 h of the filtration process to notice the stability of the extracted bath at acidic pH, and no major noticeable change was reported.

Again, the dyes can show resonating form and give different tones with the change in pH because for natural dyes, pH changes very often.

Furthermore, silk dyeing is recommended to be carried out in acidic medium as silk is sensitive to alkaline medium of dyeing, but extraction of the mangiferin dyes was optimized at alkaline pH. Therefore, the stability of the dyes after extraction is of importance.

3.3 Optimization of aqueous extraction conditions

The levels for each of the three factors in the orthogonal design of experiments are shown in Table 3. The extraction experiments were performed under optimum pH condition (pH = 10). The results of the orthogonal design of experiments are shown in Table 4.

Table 3 Factors and levels in orthogonal design of experiment for aqueous extraction

Levels	Factors		
	Temperature (°C)	Time (min)	Material-to-liquor ratio
1	80	45	1:10
2	90	60	1:13
3	98	75	1:16

Optimum factors: $A_3B_2C_1$,

Absorbance = 1.041.

The optimum extraction conditions were 98 °C for temperature factor, 60 min for time factor, and 1:10 for material-to-liquor ratio. It has been found that dye liquor extracted under optimum conditions had the maximum absorbance value, which was 1.041. From the range analysis of the average absorbance results as shown in Table 4, the most influential factor of extraction was material-to-liquor ratio, while extraction time factor was the least influential.

Table 4 The absorbance values and range analysis from orthogonal experiment

No.	A	B	C	Absorbance at 380 nm
1.	1	1	1	0.802
2	1	2	2	0.805
3.	1	3	3	0.779
4.	2	1	2	0.903
5.	2	2	3	0.859
6.	2	3	1	0.94
7.	3	1	3	0.856
8.	3	2	1	1.041
9.	3	3	2	0.999
	A	B	C	
K_1	0.795	0.849	0.988	
K_2	0.910	0.917	0.906	
K_3	0.981	0.903	0.789	
R	0.185	0.014	0.199	

K_1, K_2, K_3—the average value of each factor when its level is 1, 2, or 3. R—the difference between the maximum and minimum results for each factor

3.4 Dyed samples

The use of mordants and their combinations produced different shades on silk fabric which are shown in Table 5.

3.5 Color measurements of dyed fabrics

The results of color measurements of the dyed silks are shown in Table 6.

Table 5 Shades of dyed silk

Mordant	No mordant (Reference dyed)	Ferrous sulfate	Ferrous sulfate-alum
Shade			
	New Wheat	Desert palm	Butternut
Mordant	Ferrous sulfate-alum-tin	Alum	Alum + Tin
Shade			
	Curry	Tinsel	Cocoon
Mordant	Tin	Tin + CT	Alum-Tin-TA
Shade		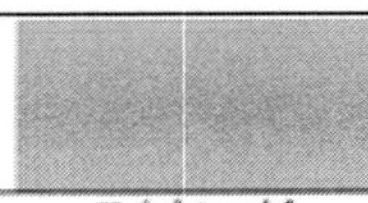	
	Straw	Bright gold	Honey mustard

Table 6 Color yield, color coordinates, and color difference of dyed fabrics

Mordant type	K/S at λ_{max}	Color strength (%)	Color coordinates					ΔE_{CMC}
			L^*	a^*	b^*	C^*	h^0	
No mordant (Reference dyed)	11.85	100	73	4.2	26.1	26.4	80.9	-
Ferrous sulfate	17.46	147.4	30.6	3.3	8.1	8.7	67.8	19
Ferrous sulfate + Alum	16.62	140.3	41.6	3	18.6	18.8	80.8	12.7
Ferrous sulfate + Alum + Tin	14.42	121.7	61.1	4.7	22.2	22.7	77.7	5.1
Alum	13.42	113.3	60.8	5.1	31.8	32.3	80.9	5.6
Alum + Tin	13.13	110.8	74.7	3	27.2	27.4	83.7	1.6
Tin	11.16	94.1	75.1	3.2	27.4	27.6	83.2	1.5
Tin + CT	13.19	111.3	65.3	5.6	32.1	32.6	80.1	4.4
Alum + Tin + TA	14.62	123.4	62.8	5.8	30.96	31.51	79.34	4.78

K/S value of the unmordanted dyed sample was found to be 11.85. This dye uptake on the silk fiber is attributed to the structural features of the fiber. However, in the mordanting method, mordant resulted in improved color yield of the dyed fabrics, except tin. Ferrous sulfate as a mordant significantly increased the color yield of silk. The K/S value was found to be 17.46 using ferrous sulfate which showed the maximum relative surface color strength value of 147.4 % considering the unmordanted dyed sample as reference. Besides, using alum with ferrous sulfate, and tin and alum with ferrous sulfate as a combination, color strengths were found to be 140.3 % ($K/S = 16.62$) and 121.7 % ($K/S = 14.42$), respectively.

In single mordanting process of silk, the order of color yield was found to be ferrous sulfate > alum > tin. It was obvious that color yield gradually decreased when approached from ferrous sulfate to tin. Again, among the four different combinations of mordants, the order was found to be ferrous sulfate-alum > alum-tin-tannic acid > ferrous sulfate-alum-tin > alum-tin.

The addition of ferrous sulfate mordant increased the greenness quality 21.58 % when compared with the reference dyed sample. Tin reduced 22.23 % redness while tin-CT increased 34.05 % redness of the reference dyed sample. Again, from b^* values, it was noticed that all the ferrous sulfate mordanted samples were bluer than the reference samples while tin-CT and alum-tin-TA mordanted samples increased yellowness of dyed fabric. The color saturation value (C^*) were found to be least in ferrous sulfate mordanted sample (8.7) whereas the values were found to be maximum in the case of alum (32.3) and tin-CT (32.6) mordanted samples. Moreover, the hue angles lie within 67.8° to 83.7°, so all of the dyed samples were closer to yellowish shade than the reddish. Higher color difference (ΔE_{CMC}) was noticeable between reference and ferrous sulfate mordanted samples, and the difference reduced from ferrous sulfate to alum and then alum to tin as shown in Table 6.

The presence of hydroxyl or carbonyl groups in dye structure is capable to form metal complex with the positively charged metals. Dye anions and metal cations have strong attraction towards positively charged amino and negatively charged carboxyl groups of silk, respectively. Hence, they form ionic bonding between dye and fiber, metal and fiber, and finally dye and metal ions. The dye-metal complex also forms coordinate bonds with the uncharged amine ($-NH_2$) groups of silk as shown in Fig. 4. In addition, one molecule of dye can form a bond with one site of fiber molecule while one molecule of mordant can form bonds with two or more molecules of dyes. Therefore, these are some of the different features indicating application of mordants increased the color yield (Bhattacharya and Shah 2000; Temani et al. 2011; Uddin 2014).

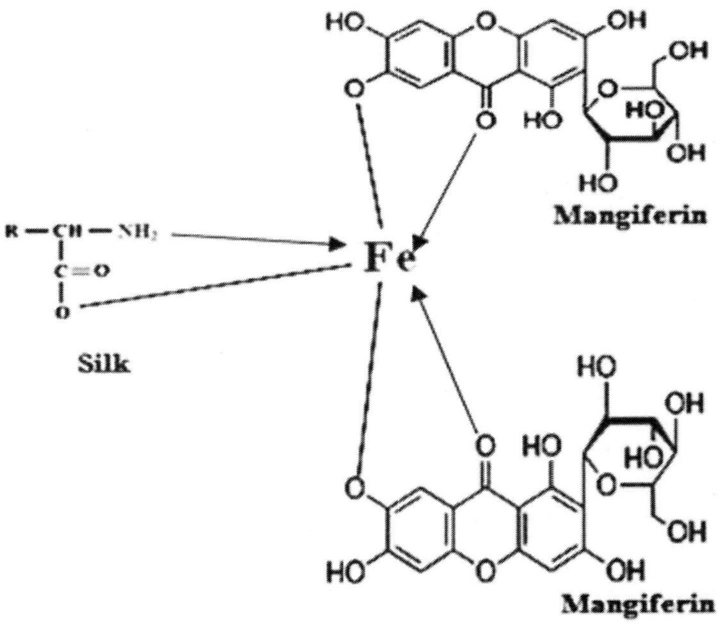

Figure. 4 Structure of mangiferin with ferrous sulfate on silk

Again, ferrous sulfate as a transition metal having coordination number 6 forms a large number of complexes with the dye molecules (Mongkholrattanasit and Punrattanasin 2012). As a result, when they interact with the silk fiber, some coordination sites remain free, and at that time, amino and carboxylic groups on the fiber can occupy these free sites. Thus, ferrous sulfate can form a ternary complex on one site with the fiber and in another site with the dye (Fig. 4). This strong coordination tendency can enhance interaction between the fiber and the dye (Bhattacharya and Shah 2000). This resulted in higher dye uptake as well as shade change due to mordanting with ferrous sulfate (Uddin2014).

In contrast, aluminum and tin salts formed weak coordination complexes with the dyes. This tends to form quite strong bonds with the dye molecule but not with the fiber (Cotton and Wilkinson 1972). Thus, they block the dye and reduce its interaction with the fiber. This is the reason behind the lower *K/S* values in the case of aluminum and tin salts than those obtained from ferrous sulfate. Moreover, CT as an assistant increased the color yield from 11.16 to 13.19 when used with tin. CT is chemically potassium hydrogen tartrate which can be used in addition to dyes and mordants to change the pH in order to change colors and to help the absorption of the mordant metal (Mortazavi et al.2012).

3.6 Fastness results

3.6.1 Washing fastness

The results of washing and light fastness of the dyed fabrics are shown in Table 7.

Table 7 Washing and light fastness of dyed fabrics

Mordant type	Washing fastness							Light fastness
	Color change	Color staining						
		Di-Acetate	Bleached cotton	Polyamide	Polyester	Acrylic	Wool	
No mordant (Reference dyed)	4	4/5	5	5	5	4/5	5	6
Ferrous sulfate	4	5	4/5	5	5	5	4	7
Ferrous sulfate + Alum	3/4	5	4/5	5	5	5	4	7
Ferrous sulfate + Alum + Tin	4	5	5	5	5	5	4	7
Alum	4/5	4/5	5	5	5	5	5	6
Alum + Tin	4/5	5	5	5	5	5	4/5	5
Tin	4/5	5	5	4/5	5	4/5	5	5
Tin + CT	5	4/5	5	4/5	5	4/5	5	7
Alum + Tin + TA	4	4/5	5	4/5	5	4/5	5	6

The unmordated dyed silk showed color change rating of 4. This can be explained that the good fastness to washing for the sample dyed without mordant was due to the affinity of coloring component through H-bonding and van der Waals forces. Using mordants, the color change ratings were found to be within 3/4 to 5, where a rating of 5 (excellent) was found using tin-CT mordant. The ratings were found to be 4/5 in the case of using alum and alum-tin. So it can be said that the overall ratings of color change were good. As wash fastness is influenced by the rate of diffusion of dye molecules and state of dyes inside the fiber, dyes has a tendency to aggregate inside the fiber. Thus, their molecular size is increased resulting in good wash fastness. In addition, in the case of mordanted samples, complexing with mordant also has the effect of insolubilizing the dye, making it color fast.

On the other hand, the color staining ratings were found to be from 4/5 to 5 for all the dyed fabrics, except when ferrous sulfate and its combinations were used as mordant. There were very slight staining observed on to the adjacent wool fiber of the multifiber fabric in the case of ferrous sulfate and its combination samples where the ratings were 4 and almost no staining on the other fibers of the multifiber fabric.

3.6.2 Light fastness

Light fastness as shown in Table 7 was found to be better, and among those, the lowest ratings attained were 5 in the case of tin and alum-tin combination while the unmordanted dyed fabric showed a rating of 6.

In the case of metallic mordants, ferrous sulfate mordanted samples dyed with the mango leaf extracts showed excellent light fastness. This happened due to the formation of a complex with transition metal which protected the chromophore from photolytic degradation. The photons sorbed by the chromophoric group dissipated their energy by resonating within the six-member ring thus formed and, hence, protecting the dyes. Thus, ferrous sulfate can bind with more dye molecules than alum or tin. During exposure to light, the fabrics mordanted with ferrous sulfate, alum, or tin may have the same number of dye molecules destroyed. But as the fabrics mordanted with ferrous sulfate had deeper shades due to bonding with more number of dye molecules, it seemed to fade less compared to the fabric mordanted with alum or tin.

4. CONCLUSIONS

This study was planned in search of greener alternative to satisfy the consumers' growing demand of eco-friendly products, and progress has been made with this study in the use of mango leaves extracts. The maximum relative color strength of the extracted dye liquor was found to be at pH 10. But the extracted dye liquors have shown good pH stability in acidic conditions. It was shown that different fashion hues were obtained on silk fabric from the same dye extract using mordants and their combinations. Again, color yields were found to be influenced by the addition of mordants. In single mordanting, the order of color yield was ferrous sulfate > alum > tin. In combined mordanting, the order was ferrous sulfate-alum > alum-tin-tannic acid > ferrous sulfate-alum-tin > alum-tin. Other color values were also found to be influenced due to mordanting. Washing and light fastness properties were found to be from good to excellent in most of the cases. Thus, on the basis of the results, it can be said that mango leaves have good scope for application on silk fabrics.

COMPETING INTERESTS

The author declares that he has no competing interest.

REFERENCES

1 Ali, S. (2007). *Evaluation of cotton dyeing with aqueous extracts of natural dyes from indigenous plants (PhD Thesis)* (pp. 62–63). Pakistan: University of Agriculture, Faisalabad. 96.

2 Ali, MA, Almahy, HA, & band Ali, AA. (2013). Extraction of carotenoids as natural dyes from the *Daucus carota* Linn (carrot) using ultrasound in Kingdom of Saudi Arabia. *Research Journal of Chemical Sciences, 3*(1), 63–66.

3 Ammayappan, L, Shakyawar, DB, Chhagan, L, Sharma, M, & Wani, SA. (2014). Extraction of natural colourants from agricultural residues

and their application on woolen fabric: part 2. *Man Made Textiles In India, 57*(6), 212–215.

4 Bains, S, Singh, OP, Goraya, G, & Kang, M. (2003). Dyeing of cotton with mango (*Mangifera indica*) dye. *Journal of the Textile Association, 64*(1), 23–26.

5 Bhattacharya, SD, & Shah, AK. (2000). Metal ion effect on dyeing of wool fabric with catechu. *Color Technol, 116*(1), 10–12.

6 Bruna, CV, & Maria, AMM. (2013). Azo dyes: characterization and toxicity—a review. *Textiles and Light Industrial Science and Technology, 2*(2), 85–103.

7 Cai, Z, Jiang, G, & Yang, S. (2001). Chemical finishing of silk fabric. *Color Technol, 117*, 161.

8 Clariant manual. Recommendations for dyeing silk. Sandoz Chemicals Ltd., Switzerland, ID 05543.00.94.

9 Cotton, FA, & Wilkinson, G. (1972). *Advanced inorganic chemistry: a comprehensive text* (3rd ed.). New York: John Wiley & Sons.

10 El-Nagar, K, Sanad, SH, Mohamed, AS, & Ramadan, A. (2005). Mechanical properties and stability to light exposure for dyed Egyptian cotton fabric with natural and synthetic dyes. *Polymer-plastics Technology and Engineering, 44*(7), 1269–1279.

11 Goodarzian, H, & Ekrami, E. (2010). Extraction of dye from madder plant (*Rubia tinctorium*) and dyeing of wool. *World Applied Sciences Journal, 9*(4), 434–436.

12 Hunger, K. (2003). *Industrial dyes*. Darmstadt: WILEY-VCH Verlag GmbH & Co. KGaA.

13 Iqbal, J, Bhatti, IA, & Adeel, S. (2008). Effect of UV radiation on dyeing of cotton fabric with extracts of henna leaves. *Indian Journal of Fiber & Textile Research, 33*, 157–162.

14 Jothi, D. (2008). Extraction of natural dyes from African marigold flowers (*Tagetes ereectal*) for textile coloration. *AUTEX Research Journal, 8*(2), 49.

15 Khan, TA, Sharma, S, & Ali, I. (2011). Adsorption of Rhodamine B dye from aqueous solution onto acid activated mango (*Magnifera indica*) leaf powder: equilibrium, kinetic and thermodynamic studies. *Journal of Toxicology and Environmental Health Sciences, 3*(10), 286–297.

16 Klaichoi, C, & Padungtos, W. (2010). Development of batik painting technique silk fabric via natural dyes. In *The 2nd RMUTP International Conference, Green Technology and Productivity* (pp. 382–387).

17 Luo, F, Lv, Q, Zhao, Y, Hu, G, Huang, G, Zhang, J, et al. (2012). Quantification and purification of mangiferin from Chinese mango (*Mangifera indica* L.) cultivars and its protective effect on human umbilical vein endothelial cells under H_2O_2-induced stress. *Int J Mol Sci, 13*, 11260–11261.

18 Mahajan, S, Sidhu, SP, & Grewal, J. (2005). Dyeing of silk with peach (*Prunus persica*) dye using combination of mordants. *Journal of the Textile Association, 66*(2), 85–89.

19 Mcdonald, R (Ed.). (1997). *Color physics for industry (2nd ed.). Soc.* Bradford: Dyers and Colorists.

20 Mongkholrattanasit, R, & Punrattanasin, N. (2012). *Properties of silk fabric dyed with eucaliptus, quercetin, rutin and tannin using padding techniques.* Bangkok, Thailand: RMUTP International Conference on Textiles & Fashion.

21 Mortazavi, SM, Kamali Moghaddam, M, Safi, S, & Salehi, R. (2012). Saffron petals, a by-product for dyeing of wool fibers. *Prog Color Colorants Coat, 5*, 75–84.

22 Rawat, B, Jahan, S, Grover, E, & Yadav, S. (2006). Color fastness properties of silk fabric dyed with poinsettia leaves. *Asian Textile Journal, 15*(4), 43–45.

23 Saravanan, P, Chandramohan, G, Mariajancyrani, J, & Shanmugasundaram, P. (2013). A study on extraction and application of eco-friendly natural dye extracted from leaves of *Acalypha indica* Linn on silk fabric. *International Journal of Textile and Fashion Technology, 3*(5), 1–8.

24 Sivakumar, V, Jayapriya, J, Shriram, V, Srinandini, P, & Swaminathan, G. (2009a). Ultrasound assisted enhancement in wattle bark (*Acacia mollissima*) vegetable tannin extraction for leather processing. *J Am Leather Chem Assoc, 104*(11), 375–383.

25 Sivakumar, V, Lakshmi, A, Vijayeeswaree, J, & Swaminathan, G. (2009b). Ultrasound assisted enhancement in natural dye extraction from beetroot for industrial applications and natural dyeing of leather. *UltrasonSonochem, 16*(6), 782–789.

26 Spyroudis, S. (2000). Hydroxyquinones: synthesis and reactivity. *Molecules, 5*, 1291–1330.

27 Suneeta, MB, & Mahale, G. (2002). Dye from Parthenium leaves. Man-made Textiles in India, 45(5), 198–200.

28 Temani, P, Shakyawar, DB, Ammayappan, L, Goyal, V, & Wani, SA. (2011). Standardization of dyeing condition of cochineal extract on pashmina yarn. *Journal of the Textile Association, 72*(2), 90–92.

29 Uddin, MG. (2014). Effects of different mordants on silk fabric dyed with onion outer skin extracts. *Journal of Textiles, Article ID 405626*, 1–8.

30 Win, ZM, & Swe, MM. (2008). Purification of the natural dyestuff extracted from mango bark for the application on protein fibers. *World Acad Sci Eng Technol, 22*, 536.

CHAPTER 8

Eco-Friendly Disperse Dyeing and Functional Finishing of Nylon 6 Using Supercritical Carbon Dioxide

Tarek Abou Elmaaty [1,*], Eman Abd El-Aziz [1], Jaehuyk Ma [2], Fathy El-Taweel [3] and Satoko Okubayashi [2]

[1] Department of Textile Dyeing & Finishing, Faculty of Applied Arts, Damietta University, Damietta 34512, Egypt

[2] Department of Advanced Fibro Science, Kyoto Institute of Technology, Kyoto 606-8585, Japan

[3] Department of Chemistry, Faculty of science, Damietta University, Damietta 34512, Egypt

ABSTRACT

In this work, a supercritical carbon dioxide assembly was successfully constructed for dyeing Nylon6 fabric. Primary experiments were carried out to confirm the possibility of bringing the dyeing up to factory scale. A series of disperse azo dyes with potential antibacterial activity were applied to dye the fabric under our study in supercritical carbon dioxide ($scCO_2$). The factors affecting the dyeing conditions (*i.e.*, dye concentration, time, temperature and pressure) and functional properties were discussed and compared with those in aqueous dyeing. The comparison revealed that elimination of auxiliary chemicals such as salt, carrier or dispersing agent has no diverse effect on dyeing. The color strength of the dyed fabric evaluated by using K/S measurements increased by increasing dye concentration from 2% to 6% owf. (on weight of fabric). The nylon6 fabrics dyed in supercritical carbon dioxide have good fastness properties, and especially light fastness compared with conventional exhaustion dyeing. Antibacterial activity of the dyed samples under supercritical conditions was evaluated and the results showed excellent antibacterial efficiency.

KEYWORDS

eco-friendly disperse dyeing; polyamide 6 fabric; supercritical carbon dioxide; antimicrobial disperse dyeing; combined process

1. INTRODUCTION

Using supercritical carbon dioxide (*scCO$_2$*) instead of water in textile dyeing can preserve energy, lower water use and prevent pollution. This dyeing method offers many advantages compared with conventional aqueous dyeing; no carrier or dispersing agent is required, residual dyestuff can be collected and carbon dioxide can be recycled [1,2,3]. It is an environmentally friendly technique, as it may replace the traditional wet-dyeing method [4].

Polyamide fibers have particular properties such as high tensile strength, elasticity and good mechanical and chemical resistance, *etc.* Correspondingly, polyamide fabrics are utilized in considerable applications of the textile industry [5,6].

The dyeability of synthetic hydrophobic fibers such as polyamide, polyester, polyacrylonitrile and polypropylene in*scCO$_2$* poses a great challenge to dyestuff chemists. As a consequence in the last few decades, various researchers focused their efforts on the synthesis of new dyes for these fibers [7,8,9,10,11,12]. The success of dyeing synthetic textiles, particularly polyester [13,14,15,16,17,18, 19,20, 21,22,23], with disperse dyes in *$scCO_2$* prompted research application of this technique to other synthetic fibers such as polypropylene [24], and aramid fibers [25] or alternative natural fabrics like cotton [26,27,28].

However, only a limited number of studies have been published on dyeing of polyamide textiles using *$scCO_2$*. The dyeing of nylon 6-6 with hydrophobic-reactive and disperse-reactive dyes using supercritical carbon dioxide as a solvent was reported [29,30]; a covalent force was formed successfully between the terminal amine group of nylon6 6 and the vinylsulphone group of the dye molecule. The work indicated that both solubility and affinity have an effect on the dye uptake of nylon 6-6 with hydrophobic reactive and disperse-reactive dyes using supercritical carbon dioxide as a solvent system. Light fastness was acceptable for common applications and washing fastness was superior.

In an earlier study [10], we explored the dyeing of polyester fabrics with antibacterial disperse–azo dyestuffs which were synthesized in our program, employing a supercritical carbon dioxide dyeing technique. Working with antibacterial dyes in textiles integrated the dyeing and finishing process and resulted in a more effective technique in terms of water and energy management. The obtained result showed that this process was absolutely as adequately efficient as the typical procedure and led us to study the behavior of the synthesized dyes on other synthetic fabrics. In this context, the purpose of this work was to provide a one-step dyeing and finishing process for nylon 6 fabrics with antimicrobial disperse dyes through supercritical processing.

2. EXPERIMENTAL SECTION

2.1. Fabric and Dyes

A 100% polyamide 6 plain plane weave fabric (70 g/m^2) supplied by Shikisen-sha company (Osaka, Japan) was used as dyeing substrate. Figure 1 shows the chemical structure of dyes employed in our research and was prepared according to the literature [31].

dye	Name (IUPAC)
1	2-(1,5-dimethyl-3-oxo-2-phenyl-2,3-dihydro-1H-pyrazol-4-yl)-2-oxo-N'-(p-tolyl) acetohydrazonoyl cyanide
2	2-(1,5-dimethyl-3-oxo-2-phenyl-2,3-dihydro-1H-pyrazol-4-yl)-N'-(2-methoxyphenyl)-2-oxoacetohydrazonoyl cyanide
3	2-([1,1'-biphenyl]-4-yl)-N'-(1,5-dimethyl-3-oxo-2-phenyl-2,3-dihydro-1H-pyrazol-4-yl)-2-oxoacetohydrazonoyl cyanide
4	N'-(2-chloro-4-methylphenyl)-2-oxo-2-(p-tolyl) acetohydrazonoyl cyanide
5	N'-(2-chlorophenyl)-2-oxo-2-(p-tolyl) acetohydrazonoyl cyanide

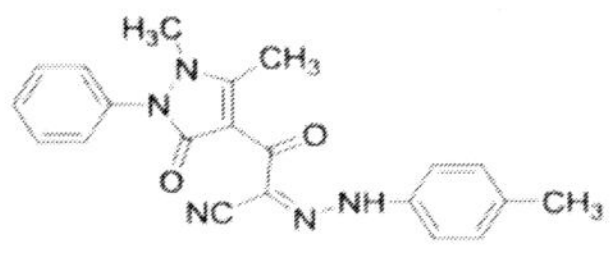

1

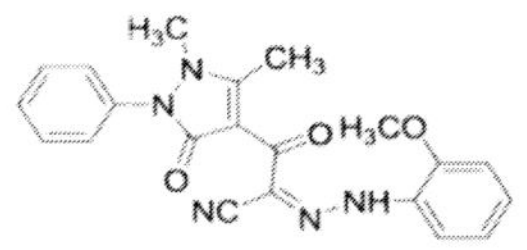

2

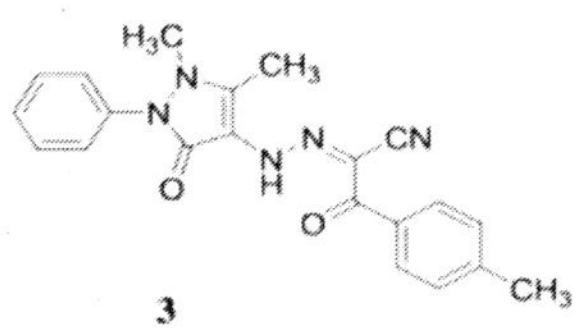

3

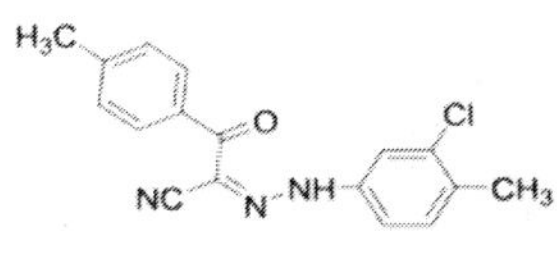

4

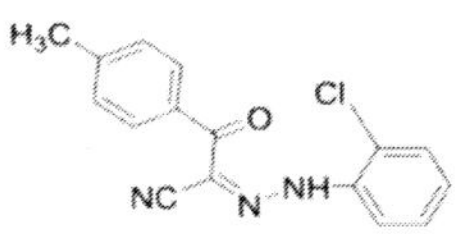

5

Figure 1. The chemical structure of dyes.

2.2. Dyeing Apparatus

Figure 2 is a diagram of the whole apparatus. The liquefied CO_2 departing from the cylinder moved inward to a cooling unit and was infused into a high-pressure syringe pump (model Jasco Pll-2880 plus, Jasco, Easton, PA, USA). High-pressure CO_2 ultimately ran out into a dyeing autoclave. The dyeing autoclave (Jasco EV-3, Jasco, Easton, PA, USA) as shown in Figure 2 is a 50 cm^3 stainless steel autoclave outfitted with a steel screw-tube, a pressure sealed magnetic stirrer, and a quick-release cap.

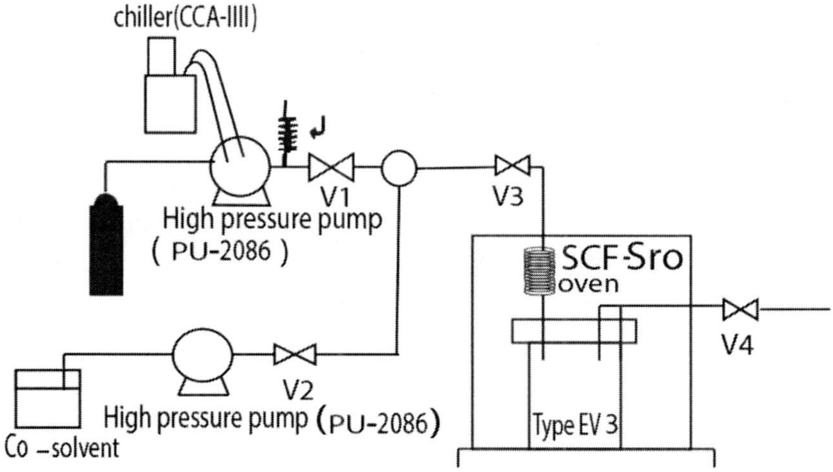

Figure 2. Supercritical CO_2 apparatus.

2.3. Procedures

2.3.1. *ScCO₂* Dyeing

Polyamide 6 fabric (usually 3 × 10 cm) was wrapped around a stainless steel cylinder coil bearing perforated holes (0.5 cm diameter) and seated inside the autoclave. The purified dye was loaded on the base of the surface of the cylinder, and the amount of dye used varied from 2% to 6% owf. The autoclave was then sealed and heated to the desired temperature. At the same time, CO_2 was pumped through into the vessel and kept at a working pressure by stirring. The head temperature of the pump was maintained at −5 °C using a chiller. The circulation system was activated as the pressure reaches 10 MPa. The stream of the fluid was introduced using the magnetic drive under the column at 750 rpm. The fluid flowed from the inside to the outside of the cylinder. After a definite reaction time (1 hand 3 h), the CO_2 released by shutting off the valve slowly until the pressure of the dyeing vessel reached atmospheric pressure. After dyeing, the fiber was removed, soaped at a temperature of 60 °C for 15 min, and then rinsed with water.

2.3.2. Aqueous Dyeing

As shown in Figure 3, the dye bath (1:20 liquor ratio) containing 5 g/dm^3 carrier, 4% ammonium sulphate was adjusted to pH 5.5 and brought to 60 °C. The polyamide 6 fabric was added at this temperature and run for 15 min. 2.0% to 6.0% (owf.) of the dyes under our study were dissolved in a solution of (2 g/dm^3), an anionic dispersing agent, followed by the dye being precipitated in a fine dispersion. The fine dispersion was then added, the temperature was raised to the boiling point over a period of 45 min, and dyeing was continued at the boiling point for about one hour. After dyeing, the samples were soaped with a detergent and some NaOH in a bath containing 2% nonionic detergent at a temperature of 60 °C for 15 min, then rinsed in water and dried at room temperature.

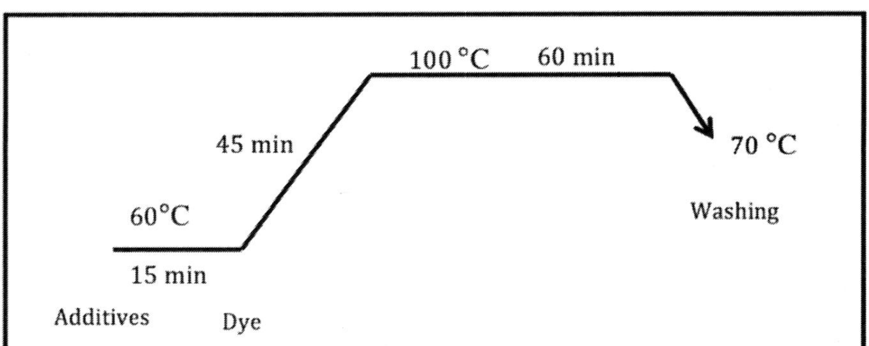

Figure 3. Exhaustion dyeing curve.

2.4. Measurements

Color strength (K/S) values of the dyed polyamide 6 fabrics were evaluated using the (Konica Minolta spectrophotometer CM-3600 d) spectrophotometer (Minolta, Tokyo, Japan).

Fastness properties, mainly washing, rubbing, and light fastness, of the dyed polyamide 6 fabrics were evaluated according to JIS L 0844, JIS L 0849, and JIS L 0842:2004 test methods, respectively [32,33,34].

The color parameters of the dyed polyamide fabric were measured using the (Konica Minolta spectrophotometer CM-3600 d) spectrophotometer. The following CIELAB coordinates were measured: lightness (L*), chroma (C*), hue (h), the degree of redness (+ve) and greenness (−ve) (a*), and the degree of yellowness (+ve) and blueness (−ve) (b*).

The antibacterial activity assessment on G+ve bacteria (Staphylococcus aureus and Bacillus subtilis) and G−ve bacteria (Escherichia coli and Pseudomonas aeruginosa) was conducted qualitatively according to the AATCC Test Method (147-1988) and expressed as zone of growth inhibition ZI (mm).

2.5. Statistical Analysis

All tests have been performed by taking the average of three sample readings. The standard error of the mean was calculated according to the equation given below and found to be +(−) 0.1.

$$SE_{\bar{X}} = \frac{S}{\sqrt{n}} \quad (1)$$

where S = sample standard deviation, n = number of observations of the sample.

3. RESULTS AND DISCUSSION

The main task of the current work is to introduce a one-step procedure for producing polyamide 6 fabric with antimicrobial functionality under supercritical carbon dioxide medium. The effect of dyeing parameters such as dye type, concentration, temperature, time and pressure as well as a comparison of the supercritical dyeing method with traditional aqueous dyeing have been investigated. The results obtained, along with appropriate discussion, are presented below.

3.1. Dyeing Properties of Hydazonopropanenitrile Dyes

3.1.1. Effect of Dye Concentration

Figure 4 shows that at the same dyeing conditions (120 °C, 15 MPa, 60 min), the color strength of nylon 6 fabric increased by increasing the dye concentration from 2% to 6% owf., but the increment in color depth becomes smaller when the concentration surpasses 4% for dyes number 2, 4 and 5. This may be attributed to the fact that these dyes have strong saturation at low concentration.

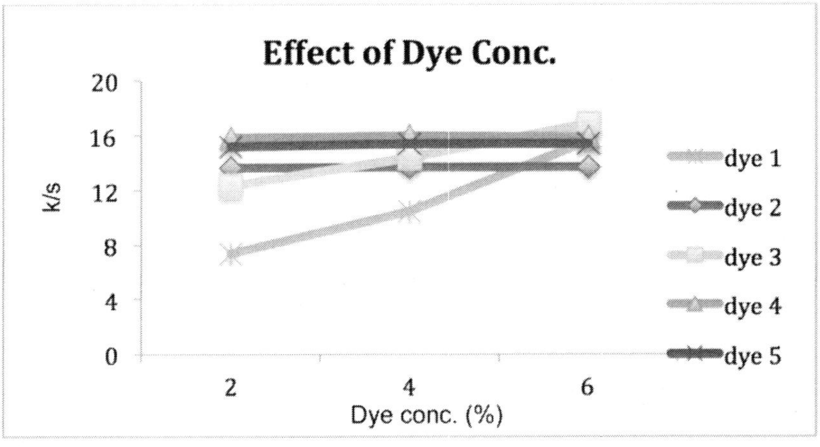

Figure 4. Effect of dye concentration.

A comparison of color strength of the $scCO_2$ and the aqueous dyed fabrics is shown in Figure 5. It was indicated that, without adding salt, carrier or dispersing agent, the appreciable color strength (K/S) of the samples dyed in $scCO_2$ was superior to those dyed in water. It can be observed that in Figure 5a, the sample dyed with dye 1, 2 and 3 in $scCO_2$ with 2% conc. has a higher K/S value than samples dyed in water with 6% dye conc. (Figure 5c). This may be attributed to the fact that dye uptake was improved by a large margin when using $scCO_2$ as a dyeing medium. For dyes 4 and 5, both have higher K/S values in supercritical conditions than in water conditions, but with a smaller border than those detected in dyes 1, 2 and 3. This means dyeing in a $scCO_2$ system exhibited significant advantages compared with traditional water dyeing. The exclusive dyeing procedure of the $scCO_2$ system is the main reason [35].

3.1.2. Effect of Dyeing Temperature

As shown in Figure 6a, the dye adsorption or uptake, characterized as color strength of K/S, remarkably increased with increasing system temperature, especially for temperatures higher than 100 °C. The significant effect of temperature could be explained by the fact that, higher system temperatures lead to higher activities of the molecules of dyestuff and supercritical carbon dioxide fluid, as well as an increase in flexibility of the nylon polymer chains. The rubbery and amorphous regions of the polymer were increased compared to the harder and more brittle areas, resulting in greater permeability of the dye molecules [4,36].

3.1.3. Effect of Dyeing Pressure

As shown in Figure 6b, the dye uptake expressed as the color strength K/S was remarkably improved with an increasing system pressure. This behavior could be made clear by the fact that increasing system pressure led to an increase in the density of supercritical carbon dioxide fluid, which consequently increased its solvent power. Hence the dyes could be readily dissolved, as well as enhancing swelling of the nylon fibers in the supercritical dyeing medium, resulting in a higher dye adsorption and enhancement in color strength value [36].

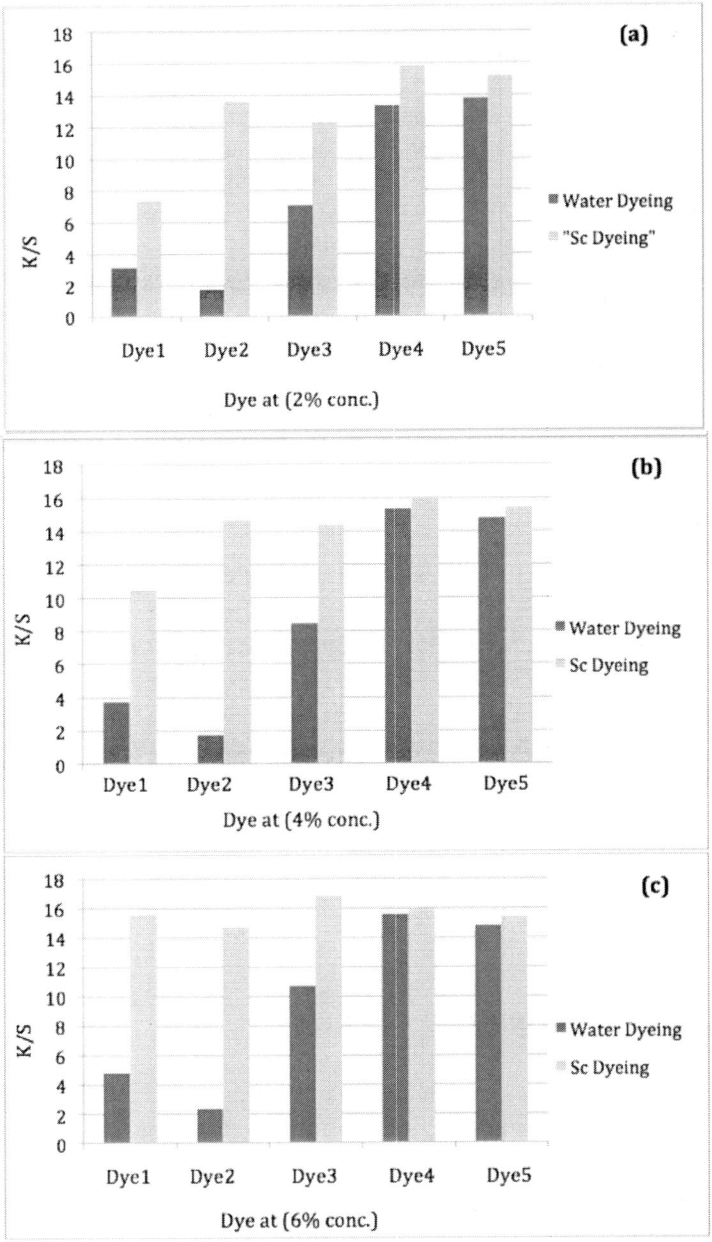

Figure 5. Comparison of color strength of *$scCO_2$* and the aqueous dyeing (**a**) at 2% dye conc; (**b**) at 4% dye conc; (**c**) at 6% dye conc.

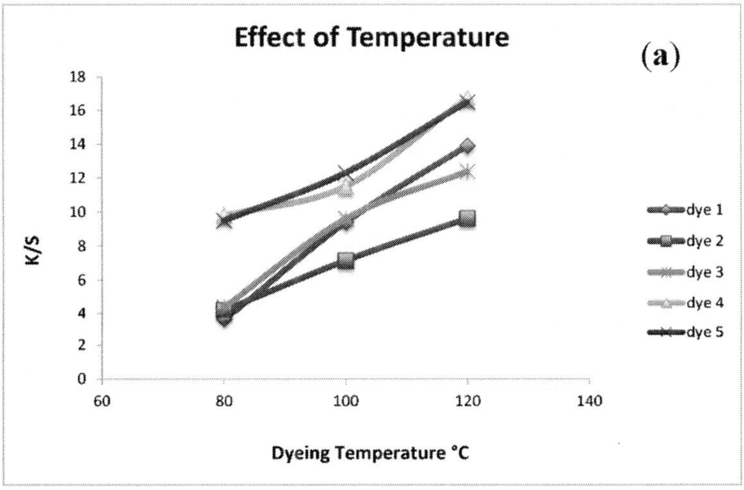

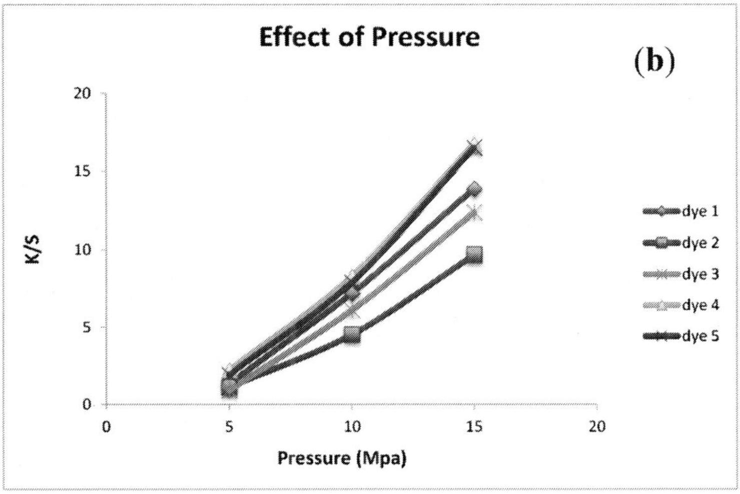

Figure 6. (**a**) Effect of dyeing temperature; (**b**) Effect of dyeing pressure.

3.1.4. Effect of Dyeing Time

The relationship between dye adsorption and dyeing time (1hand3h) in *$scCO_2$* is demonstrated in Figure 7a–c. It is seen that dye uptake expressed as color strength (K/S) increased with increasing dyeing time, at all dye concentrations (2%, 4% and 6%). The improvement in K/S values reflected the positive impact of increasing dyeing time which led to an adequate and uniform adsorption of the dye by the fibers, as well as uniform penetration and diffusion of the dye into the fabric, which resulted in the enhancement of the uptake of the dye into the fabric [36,37]. However, this result was not consistent with all dyes, since the K/S value of the fabric dyed with dye 2 for 1 h is higher than that of the fabric dyed with dye 2 for 3 h as shown in Figure 7a. This may be attributed to the decomposition of the dye with prolonged heating leading to a lower K/S value.

3.2. Color Fastness

The color fastness of the dyed nylon 6 fabrics with the proposed dyes was evaluated and recorded in Table 1. The washing fastness rating of the dyed nylon 6 fabric under supercritical medium was excellent for both fading and staining with ratings ranging from 4 to 5, while those of exhaustion dyeing were excellent for staining and ranged from moderate to excellent for fading (2–5). This result is probably due the high affinity of the colored hydrazonopropanenitrile dyes for nylon 6 fibers. Furthermore, all the dyed fabrics presented very good rubbing fastness, indicating good diffusion and penetration of the dyes under our study into fiber substrates. Nevertheless, the poor light fastness of all the water dyed fabrics was noticed. On the contrary, the dyed nylon 6 fabrics under supercritical conditions had relatively excellent light fastness (rating 4–5), which should be attributed to its higher level of dye molecule aggregation and superior depth of shade.

3.3. Antimicrobial Activity

The antimicrobial activities of the nylon fabric samples dyed in both supercritical and water media were screened using an agar-well diffusion technique against four different microbial cultures. The antimicrobial results (Table 2) attained were found to be fairly good owing to the presence of potentially active function groups, e.g., chloro, cyano and antipyrine moiety in the structure of dyes [38]. The results can be interpreted in terms of nonspecific action, *i.e.*, antibacterial activity can be achieved either by causing damage to bacterial cells or by means of restriction of a specific bacterial target

[39]. The dyeing technique has practically no effect on the imparted antimicrobial properties.

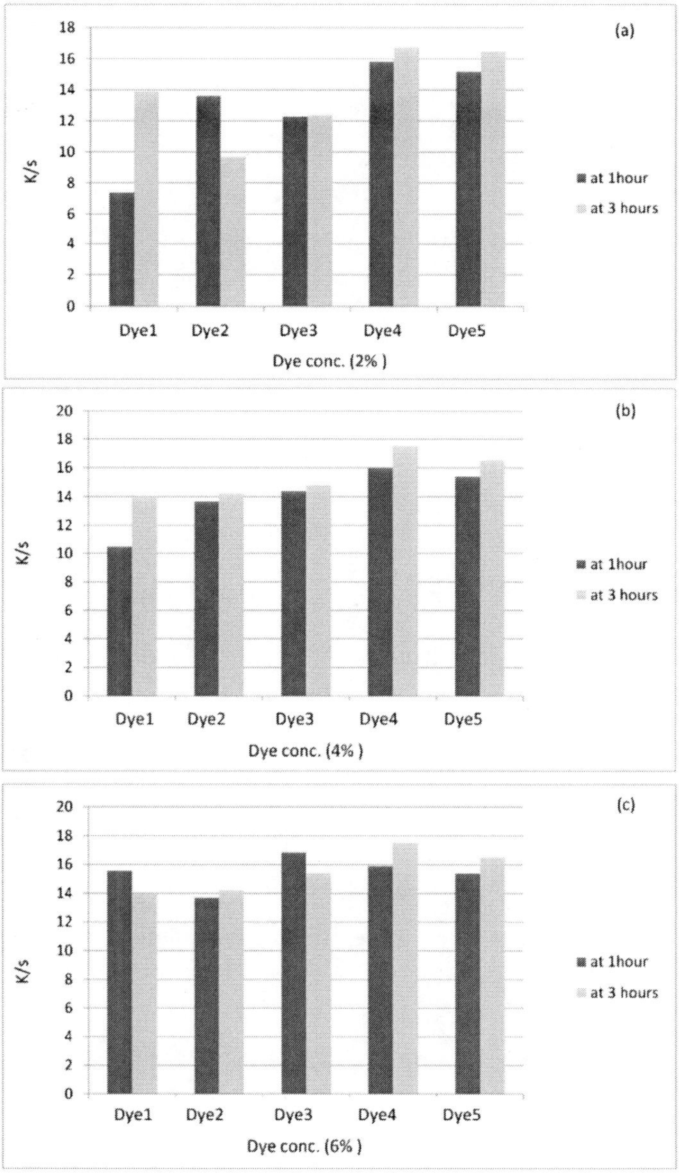

Figure 7. (**a**) Effect of dyeing time at 2% dye conc; (**b**) at 4% dye conc; (**c**) at 6% dye conc.

Table 1. Fastness properties of dyed nylon 6 samples.

Dye	$ScCO_2$ Dyeing				Aqueous Dyeing			
	Rubbing Fastness	Washing Fastness		Light Fastness	Rubbing Fastness	Washing Fastness		Light Fastness
		Color Change	Staining			Color Change	Staining	
1	5	5	5	4	5	5	5	1–2
2	4–5	5	5	4	5	5	5	1
3	4–5	5	5	4–5	2–3	5	5	1
4	4–5	4–5	5	4–5	3–4	2–3	5	1
5	5	4–5	5	5	4	2–3	4	1–2

Table 2. Antibacterial activity of dyed samples.

Dye	ZI of the Dyed Nylon in $scCO_2$				ZI of the Dyed Nylon in Water			
	G-ve		G+ve		G-ve		G+ve	
	Pseudomonas Aeruginosa	*Escherichia coli*	*Bacillussubtilis*	*Staphylococcus Aureus*	*Pseudomonasaeruginosa*	*Escherichia coli*	*Bacillussubtilis*	*Staphylococcus S. aureus*
1	12	13	13	13	12	13	12	13
2	13	12	14	13	12	13	14	13
3	14	13	14	13	13	13	13	13
4	13	12	14	13	12	12	12	13
5	12	11	12	11	11	11	12	11

ZI: zone of inhibition

3.4. Color Assessment

The color of the supercritical dyed nylon 6 fabrics was evaluated using the CIELAB system in terms of L^*, a^*, and b^* (Table 3). The color coordinates recorded in Table 3 indicated that the dye has good affinity for nylon 6 fabric and favored the following characteristics:

The dyes in our study displayed good affinity for nylon 6 fabrics at the given temperature and present generally bright and deep hues ranging from yellow to orange.

The color hues of the dyes on nylon 6 fabrics were shifted towards the yellowish direction on the yellow-blue axis according to the positive values of b^*.

The color hues of the dyes on nylon 6 fabric were shifted towards the greenish direction on the red- green axis as indicated from the negative value of a^*.

Table 3. Color coordinates of the dyed nylon samples.

Dye	*L**	*C**	*H*	*a**	*b**
1	88.11	79.1	97.84	−10.79	78.36
2	90.75	90.34	100.11	−12.4	88.94
3	92.12	76.33	102.71	−16.79	74.46
4	90.04	95.11	99.76	−14.31	83.16
5	92.73	81.86	102.51	−17.69	79.75

4. CONCLUSIONS

The conventional dyeing process using water as a solvent has drawbacks. Different agents have to be added for treatment of hydrophobic material; after dyeing, a consequent drying process with high energy consumption is imperative; and a large amount of wastewater is used. In contrast, dyeing with $scCO_2$ is water-free. The results confirm that hydrazonopropanenitrile azo dyes are convenient for dyeing nylon6 fabrics in $scCO_2$ and help expand the application of supercritical technology. The innovative supercritical model was designed for cleaner production of antimicrobial nylon6 fabrics.

ACKNOWLEDGMENTS

The authors wish to express their deep appreciation to the STDF (Science and Technological Development Fund) in Egypt and the Kyoto Institute of Technology in Japan for their help, which enabled this work to be carried out.

AUTHOR CONTRIBUTIONS

Satoko Okubayashi and Tarek Abou Elmaaty designed experiments; Fathy El-Taweel synthesized the dyes; Eman Abd El-Aziz and Jaehuyk

Ma performed the experiments; Tarek Abou Elmaaty and Eman Abd El-Aziz wrote the manuscript.

CONFLICTS OF INTEREST

The authors declare no conflict of interest.

REFERENCES

1 Bach, E.; Cleve, E.; Schollmeyer, E. Past, present and future of supercritical fluid dyeing technology—An overview.*Rev. Prog. Color.* **2002**, *32*, 88–102.

2 Banchero, M. Supercritical fluid dyeing of synthetic and natural textiles—A review. *Color. Technol.* **2012**, *129*, 2–17.

3 Montero, G.A.; Smith, C.B.; Hendrix, W.A.; Butcher, D.L. Supercritical Fluid Technology in Textile Processing: An Overview. *Ind. Eng. Chem. Res.* **2000**, *39*, 4806–4812.

4 Miah, L.; Ferdous, N.; Azad, M.M. Textiles Material Dyeing with Supercritical Carbon Dioxide (CO_2) without Using Water. *Chem. Mater. Res.* **2013**, *3*, 38–40.

5 Makhlouf, C.; Ladhari, N.; Roudesli, S.; Sakly, F. Influence of grafting with acrylic acid on the dyeing properties of polyamide 6.6 fibres. *Color. Technol.* **2012**, *128*, 176–183.

6 Bahtiyari, M.I. Laser modification of polyamide fabrics. *Opt. Laser Technol.* **2011**, *43*, 114–118.

7 Giorgi, M.R.D.; Cadoni, E.; Maricca, D.; Piras, A. Dyeing polyester fibres with disperse dyes in supercritical CO_2. *Dye. Pigment.* **2000**, *45*, 75–79.

8 Gharanjig, K.; Arami, M.; Bahrami, H.; Movassagh, B.; Mahmoodi, N.M.; Rouhani, S. Synthesis, spectral properties and application of

novel monoazo disperse dyes derived from N-ester-1,8-naphthalimide to polyester. *Dye. Pigment.* **2008**, *76*, 684–689.

9 Miyazaki, K.; Tabatab, I.; Horia, T. Relationship between colour fastness and colour strength of polypropylene fabrics dyed in supercritical carbon dioxide: Effect of chemical structure in 1, 4-bis (alkylamino)anthraquinone dyestuffs on dyeing performance. *Color. Technol.* **2011**, *128*, 60–67.

10 Abou Elmaaty, T.; Ma, J.; El-Taweel, F.; Abd El-Aziz, E.; Okubayashi, S. Facile Bifunctional Dyeing of Polyester under Supercritical Carbon Dioxide Medium with New Antibacterial Hydrazono Propanenitrile Dyes. *Ind. Eng. Chem. Res.* **2014**, *53*, 15566–15570.

11 Miyazaki, K.; Tabatab, I.; Horia, T. Effects of molecular structure on dyeing performance and colour fastness of yellow dyestuffs applied to polypropylene fibres in supercritical carbon dioxide. *Color. Technol.* **2011**, *128*, 51–59.

12 Ehrhardt, A.; Tabata, I.; Hisada, K.; Hori, T. Impregnation of hydroxy-anthraquinone compounds into polypropylene fabrics by $scCO_2$ and the sorption ability for metal ions. *Sen'i Gakkaishi* **2005**, *61*, 201–203.

13 Kima, T.K.; Sonb, Y.A.; Limc, Y.J. Affinity of disperse dyes on poly (ethylene terephthalate) in non-aqueous media: Part 1. Adsorption and solubility properties. *Dye. Pigment.* **2005**, *64*, 73–78.

14 Kima, T.K.; Son, Y.A. Affinity of disperse dyes on poly (ethylene terephthalate) in non-aqueous media. Part 2: Effect of substituents. *Dye. Pigment.* **2005**, *66*, 19–25.

15 Özcan, A.S.; Özcan, A. Adsorption behavior of a disperse dye on polyester in supercritical carbon dioxide. *J. Supercrit. Fluid.* **2005**, *35*, 133–139.

16 Banchero, M.; Ferri, A.; Manna, L. The phase partition of disperse dyes in the dyeing of polyethylene terephthalate with a supercritical CO_2/methanol mixture. *J. Supercrit. Fluid.* **2009**, *48*, 72–78.

17 Banchero, M.; Ferri, A.; Manna, L.; Sicardi, S. Dye uptake and partition ratio of disperse dyes between a PET yarn and supercritical carbon dioxide. *J. Supercrit. Fluid.* **2006**, *37*, 107–114.

18 Tabata, I.; Lyu, J.; Cho, S.; Tominaga, T.; Hori, T. Relationship between the solubility of disperse dyes and the equilibrium dye adsorption in supercritical fluid dyeing. *Color. Technol.* **2001**, *117*, 346–351.

19 Kawahara, Y.; Yoshioka, T.; Sugiura, K.; Ogawa, S.; Kikutani, T. Dyeing behavior of high-speed spun poly (ethylene terephthalate) fibers in supercritical carbon dioxide. *J. Macromol. Sci. Part B Phys.* **2001**, *40*, 189–197.

20 Bao, P.; Dai, J. Relationships between the Solubility of C.I. Disperse Red 60 and Uptake on PET in Supercritical CO_2. *J. Chem. Eng. Data* **2005**, *50*, 838–842.

21 Hou, A.; Xie, K.; Dai, J. Effect of Supercritical Carbon Dioxide Dyeing Conditions on the Chemical and Morphological Changes of Poly (ethylene terephthalate) Fibers. *J. Appl. Polym. Sci.* **2004**, *92*, 2008–2012.

22 Filho, L.C.; Mazzer, H.R.; Santos, J.C.; Andreaus, J.; Feihrmann, A.C.; Beninca, C.; Cabral, V.F.; Zanoelo, E.F. Dyeing of polyethylene terephthalate fibers with a disperse dye in supercritical carbon dioxide. *Text. Res. J.* **2014**, *84*, 1279–1287.

23 Kraan, M.V.D.; Cid, M.V.F.; Woerlee, G.F.; Veugelers, W.J.T.; Witkamp, G.J. Equilibrium Study on the Disperse Dyeing of Polyester Textile in Supercritical Carbon Dioxide. *Text. Res. J.* **2007**, *7*, 550–558.

24 Liao, S.K.; Chang, P.S.; Lin, Y.C. Analysis on the Dyeing of Polypropylene Fibers in Supercritical Carbon Dioxide. *J. Polym. Res.* **2000**, *7*, 155–159.

25 Kim, T.; Kim, G.; Park, J.Y.; Lim, J.S.; Yoo, K.P. Solubility Measurement and Dyeing Performance Evaluation of Aramid NOMEX Yarn by Dispersed Dyes in Supercritical Carbon Dioxide. *Ind. Eng. Chem. Res.* **2006**, *45*, 3425–3433.

26 Gao, D.; Yang, D.F.; Cui, H.S.; Huang, T.T.; Lin, J.X. Synthesis and Measurement of Solubilities of Reactive Disperse Dyes for Dyeing Cotton Fabrics in Supercritical Carbon Dioxide. *Ind. Eng. Chem. Res.* **2014**, *53*, 13862–13870.

27 Ozcan, A.S.; Clifford, A.A.; Bartlea, K.D.; Lewis, D.M. Dyeing of Cotton Fibres with Disperse Dyes in Supercritical Carbon Dioxide. *Dye. Pigment.* **1998**, *36*, 103–110.

28 Beltrame, P.L.; Castelli, A.; Selli, E.; Mossa, A.; Testa, G.; Bonfattic, A.M.; Seves, A. Dyeing of Cotton in Supercritical Carbon Dioxide. *Dye. Pigment.* **1998**, *39*, 335–340.

29 Liao, S.K.; Ho, Y.C.; Chang, P.S. Dyeing of nylon 66 with a disperse-reactive dye using supercritical carbon dioxide as the transport medium. *JSDC* **2000**, *116*, 403–407.

30 Liao, S.K. Dyeing Nylon-6,6 with Some Hydrophobic Reactive Dyes by Supercritical Processing. *J. Polym. Res.* **2004**,*11*, 285–291.

31 Abou Elmaaty, T.; El-Taweel, F.; Abd El-Aziz, E.; Yuesf, M.; Okubayashi, S. Facile bifunctional dyeing of polyester fabrics with new antibacterial β-oxoalkanenitriles disperse dyes. *Int. J. Sci. Eng. Res.* **2014**, *5*, 703–706.

32 *JIS L 0844: Test Methods for Color Fastness to Washing and Laundering*; Japanese Standards Association: Minato-ku, Tokyo; Suga Weathering Technology Foundation: Tokyo, Japan, 2011.

33 *JIS L 0849: Test Methods for Color Fastness to Rubbing*; Japanese Standards Association: Minato-ku, Tokyo; Suga Weathering Technology Foundation: Tokyo, Japan, 2013.

34 *JIS L 0842: Test Methods for Colour Fastness to Enclosed Carbon Arc Lamp Light*; Japanese Standards Association: Minato-ku, Tokyo; Suga Weathering Technology Foundation: Tokyo, Japan, 2004.

35 Hou, A.; Chen, B.; Dai, J.; Zhang, K. Using supercritical carbon dioxide as solvent to replace water in polyethylene terephthalate (PET) fabric dyeing procedures. *J. Clean. Prod.* **2010**, *18*, 1009–1014.

36 Long, J.J.; Ma, Y.Q.; Zhao, J.P. Investigations on the level dyeing of fabrics in supercritical carbon dioxide. *J. Supercrit. Fluid.* **2011**, *57*, 80–86.

37 Hou, A.; Dai, J. Kinetics of dyeing of polyester with CI Disperse Blue 79 in supercritical carbon dioxide. *Color. Technol.* **2005**, *121*, 18–20.

38 Elattar, K.M. Synthesis of Novel Azo Disperse dyes Derived from 4-Aminoantipyrine and their Applications to Polyester Fabrics. *Am. J. Org. Chem.* **2012**, *2*, 52–57.

39 Ibrahim, N.A.; Eid, B.M.; Abou Elmaaty, T.M.; Abd El-Aziz, E. A smart approach to add antibacterial functionality to cellulosic pigment prints. *Carbohyd. Polym.* **2013**, *94*, 612–618.

CHAPTER 9

Degradation of textile dyes using immobilized lignin peroxidase-like metalloporphines under mild experimental conditions

Paolo Zucca[1,2] Antonio Rescigno[1], Manuela Pintus[1], Andrea C Rinaldi[1] and Enrico Sanjust[1]

[1]Dipartimento di Scienze Biomediche, Università di Cagliari, Cittadella Universitaria, 09042, Monserrato, CA, Italy

[2]Consorzio UNO (University of Oristano Consortium), Oristano, Italy

ABSTRACT

Background

Synthetic dyes represent a broad and heterogeneous class of durable pollutants, that are released in large amounts by the textile industry. The ability of two immobilized metalloporphines (structurally emulating the ligninolytic peroxidases) to bleach six chosen dyes (alizarin red S, phenosafranine, xylenol orange, methylene blue, methyl green, and methyl orange) was compared to enzymatic catalysts. To achieve a green and sustainable process, very mild conditions were chosen.

Results

IPS/MnTSPP was the most promising biomimetic catalyst as it was able to effectively and quickly bleach all tested dyes. Biomimetic catalysis was fully characterized: maximum activity was centered at neutral pH, in the absence of any organic solvent, using hydrogen peroxide as the oxidant. The immobilized metalloporphine kept a large part of its activity during multi-cycle use; however, well-known redox mediators were not able to increase its catalytic activity. IPS/MnTSPP was also more promising for use in industrial applications than its enzymatic counterparts (lignin peroxidase, laccase, manganese peroxidase, and horseradish peroxidase).

Conclusions

On the whole, the conditions were very mild (standard pressure, room temperature and neutral pH, using no organic solvents, and the most environmental-friendly oxidant) and a significant bleaching and partial mineralization of the dyes was achieved in approximately 1 h. Therefore, the process was consistent with large-scale applications. The biomimetic catalyst also had more promising features than the enzymatic catalysts.

KEYWORDS

Biomimetic, Dye, Lignin-peroxidase, Porphine, Porphyrin

1. BACKGROUND

Textile dyes are a broad and heterogeneous class of molecules used in many technological fields (i.e., textile and paper production, food technology, and hair coloring) [1]. Many chemical classes of synthetic dyes are frequently employed on an industrial scale; azo dyes, triarylmethanes, phenothiazine, and anthraquinones are among the most widespread [2, 3].

Because the goal of the dyeing process is color durability, textile dyes of all chemical classes are commonly highly resistant to both chemical and physical degradation. Accordingly, textile dyes represent a serious environmental concern because of their chemical inertness, high annual production (over 10,000 tons per year [1]), toxicity/carcinogenicity [4, 5], and the large amount of dyes released in industrial wastewaters (up to 50% of total process intake [6]).

The European Directive 2002/61/EC forbids the use of some products (in particular some azo dyes). However, these restrictions only marginally solve the environmental problem because of the large volume of other dyes discharged every year.

Several physical, chemical and enzymatic approaches have been proposed to remove dyes from industrial wastewaters [7–16] that require long reaction times, extreme operational conditions, and high costs [1]. Accordingly, the development of a clean and environmentally friendly process for removing dyes is a challenging issue, far from being solved.

Redox-active metalloporphines are efficient catalysts; in the presence of a proper oxidant [17–21], they are able to oxidize many substrates, including several textile dyes, under very mild conditions. We previously studied the ability of an immobilized manganese porphine to oxidize alizarin red S (ARS), an anthraquinone dye) [22], and phenosafranine (PNS), a phenylphenazinium dye [23], elucidating several aspects of the catalytic mechanism.

In this study, we compare the catalysts IPS/MnTSPP and PP-PVA/FeTFPP (Figure 1), each containing a metalloporphine (Mn- and Fe-, respectively) immobilized onto a solid support, thus emulating ligninolytic peroxidases [24]. We investigated the ability of these two catalysts to oxidize different chemical classes of textile dyes in the presence of hydrogen peroxide under mild conditions. We also report the results of these catalysts on ARS and PNS to allow a comprehensive comparison. All the dyes included in the study are reported in Table 1.

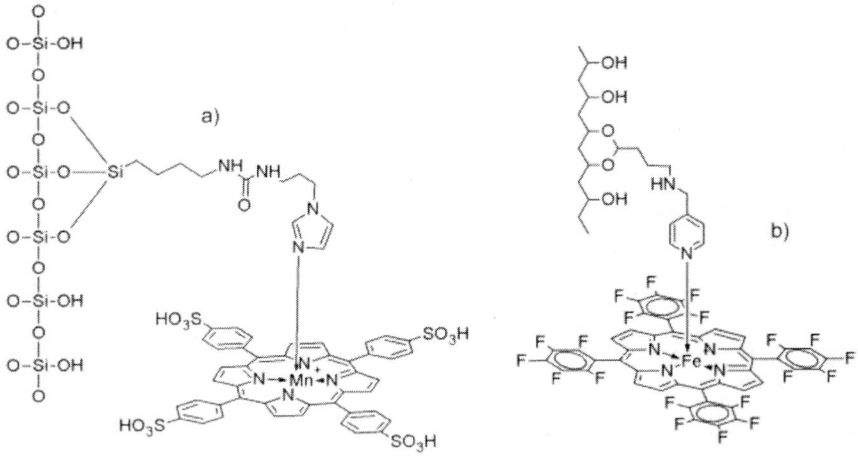

Figure 1 Proposed Structure Of The Catalysts Ips/Mntspp (A) And Pp-Pva/Fetfpp (B). For More Details, See: [24, 25].

2. RESULTS AND DISCUSSION

2.1 Comparison of IPS/MnTSPP and PP-PVA/FeTFPP

We previously developed two biomimetic catalysts, IPS/MnTSPP and PP-PVA/FeTFPP, which are able to oxidize several lignin-model compounds. IPS/MnTSPP strictly emulates a lignin peroxidase (LiP) active site, having an imidazole-grafted support that coordinates a Mn(III)-porphine [24]. This catalyst showed a LiP-like activity, whereas, unlike an enzyme such as manganese peroxidase (MnP), it was not able to use the Mn^{2+} ion as a redox mediator. Conversely, PP-PVA/FeTFPP was only "bioinspired" (i.e., only inspired by but not strictly emulating, a natural enzyme active site) because the metalloporphine was coordinated by a pyridine-grafted support [25] rather than imidazole, as in ligninolytic peroxidases. This catalyst showed both LiP-like and MnP-like catalytic activities.

Table 1 The Six Textile Dyes Included In This Study

Dye		MW (g/mol)	λ_{max} (nm)	ε (mol * L^{-1} * cm^{-1})
Alizarin Red S (ARS)	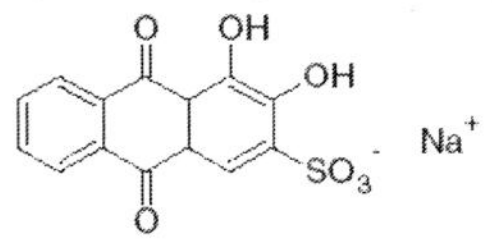	342.26	520	7200
Phenosafranine (PNS)	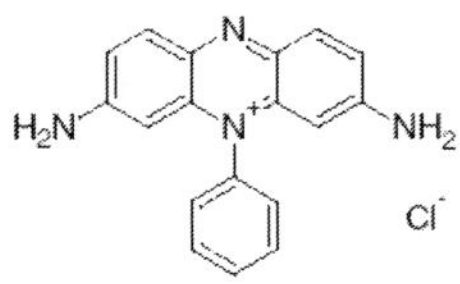	322.80	520	32000
Xylenol Orange (XO)	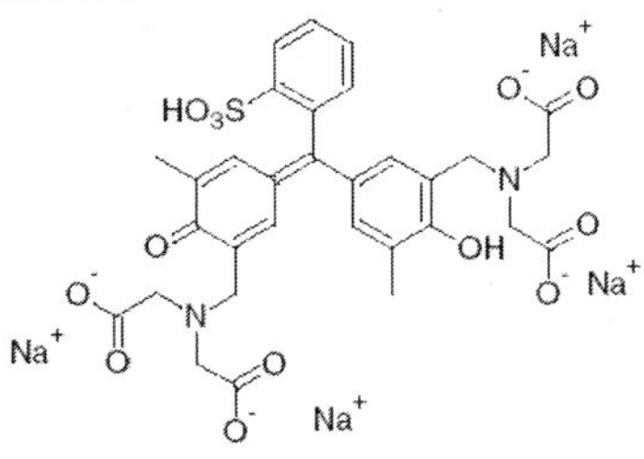	782.56	435	10500
Methylene Blue (MB)	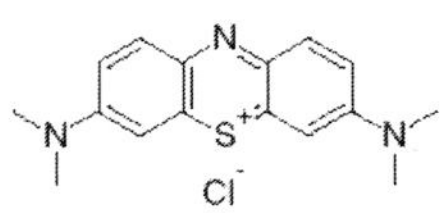	319.86	660	79500
Methyl Green (MG)	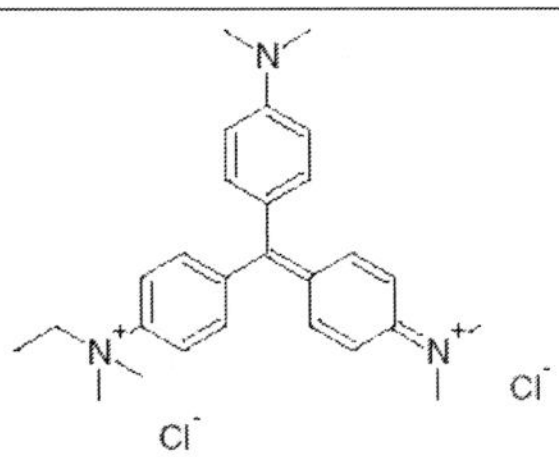	472.51	632	55700
Methyl Orange (MO)	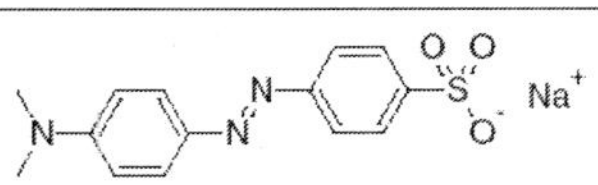	327.34	464	24800

These two catalysts were tested for their ability to decolorize several textile dyes in the presence and absence of Mn^{2+} using hydrogen peroxide as the oxidant in accordance with green chemistry principles [26]. The investigation employed very mild conditions (neutral pH and room temperature, using no organic solvents, and the most environmental-friendly oxidant) to serve as the basis for an environmentally sustainable textile wastewater treatment process.

The results of these tests are reported in Table 2. Both biomimetic catalysts easily degraded all tested dyes, showing wide substrate specificity. As previously reported [24], the catalytic behavior of IPS/MnTSPP was not affected by the presence of Mn(II), showing similar bleaching percentages in both the presence and absence of this ion. All six dyes were rapidly bleached under the conditions studied; more than with a removal in all cases higher than 50% of each dye (up to > 90% for PNS, MG, and MO) was removed in 1 h. In contrast, PP-PVA/FeTFPP was almost inactive in the absence of a suitable redox mediator. The conversion of all six dyes increased in the presence of Mn^{2+}, leading to values similar on average to those observed for IPS/MnTSPP.

Table 2 Removal of the studied dyes in the presence of hydrogen peroxide and either IPS/MnTSPP or PP-PVA/FeTFPP, in both the presence and absence of Mn(II) as the redox mediator

Mn(II)	Catalyst	Dyes removal % (1 h)					
		ARS	PNS	XO	MB	MG	MO
No	IPS/MnTSPP	62 ± 1	92 ± 2	42 ± 4	96 ± 4	82 ± 5	87 ± 2
	PP-PVA/FeTFPP	n.d.	n.d.	n.d.	n.d.	n.d.	n.d.
Yes	IPS/MnTSPP	55 ± 1	88 ± 2	44 ± 2	91 ± 3	77 ± 4	85 ± 4
	PP-PVA/FeTFPP	48 ± 3	58 ± 2	55 ± 5	37 ± 1	48 ± 2	91 ± 2

The reaction mixtures contained 5 mg of catalyst in 1 mL of 25 mM buffer solution containing 8.8 mM hydrogen peroxide and a proper concentration of the selected dyes (0.29 mM ARS, 0.31 mM PNS, 2 mM XO, 0.15 mM MB, 1.5 mM MG, and 1.25 mM MO). Quantification of the removed dye was obtained photometrically, using the λ_{max} and molar extinction coefficients reported in Table 1. n.d.: not detectable. ($n = 4$).

The reaction mixtures contained 5 mg of catalyst in 1 mL of 25 mM buffer solution containing 8.8 mM hydrogen peroxide and a proper concentration of the selected dyes (0.29 mM ARS, 0.31 mM PNS, 2 mM XO, 0.15 mM MB, 1.5 mM MG, and 1.25 mM MO). Quantification of the removed dye was obtained photometrically, using the λ_{max} and molar extinction coefficients reported in Table 1. n.d.: not detectable. ($n = 4$).

Considering that textile wastewaters do not usually contain manganous ion and that it would therefore be necessary to add Mn^{2+}, itself a heavy metal pollutant, to the reaction, IPS/MnTSPP was considered the most promising catalyst for future industrial applications Hence, the rest of study focused on IPS/MnTSPP.

The time dependence of dye bleaching by IPS/MnTSPP is shown in Figure 2. Note that most of the bleaching occurs in the first hour of the reaction, consistent with industrial timing needs.

2.2 Characterization of bleaching by IPS/MnTSPP and H_2O_2

IPS/MnTSPP catalysis was strongly pH-dependent; an approximately neutral pH was optimal for all dyes tested except XO and PNS, which had optimal bleaching at pH 6 and 8, respectively (data not shown). However, XO and PNS were sufficiently bleached at pH 7 that IPS/MnTSPP remains a viable catalyst for environmentally friendly wastewater treatment.

An exhaustive kinetic modeling of this catalysis is beyond the scope of our work. However, the decolorization of all six dyes apparently followed saturation Michaelis-Menten-like kinetics, allowing the calculation of catalytic parameters (Table 3). Reported Michaelis constants (K_M) showed that IPS/MnTSPP had significantly higher affinity for cationic dyes than anionic dyes (lower K_M for PNS and MB, higher K_M for ARS and XO). This lower affinity led to the lowest bleaching rate for XO and ARS (compare Figure 2). Notably, in all cases, K_M values were in the millimolar range and did not suffer from substrate inhibition. This can be a crucial issue for potential industrial applications because some bleaching approaches (i.e., enzymatic, *vide ultra*) are not able to work at these concentrations.

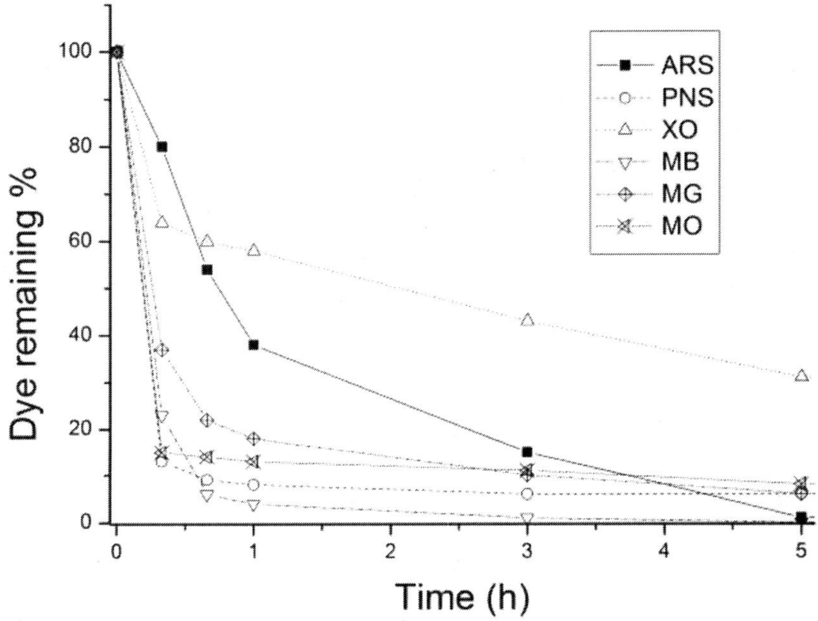

Figure 2 Time Dependence Of Dye Removal In The Presence Of Ips/Mntspp And Hydrogen Peroxide. The reaction mixtures contained 5 mg of catalyst in 1 mL of 25 mM buffer solution containing 8.8 mM hydrogen peroxide and a proper concentration of the selected dyes (0.29 mM ARS, 0.31 mM PNS, 2 mM XO, 0.15 mM MB, 1.5 mM MG, and 1.25 mM MO). Quantification of the dyes was obtained photometrically, using the λ_{max} and molar extinction coefficients reported in Table 1.

Table 3 Kinetic Parameters Measured For The Bleaching Of Each Dye By Ips/Mntspp And Hydrogen Peroxide

Sub-strate	Kinetic parameter	Dye					
		ARS	PNS	XO	MB	MG	MO
Reducing substrate	K_M (mM)	2.11 ± 0.74	0.32 ± 0.04	1.33 ± 0.38	0.11 ± 0.02	0.72 ± 0.12	0.63 ± 0.05
	k_{cat} (min^{-1})	0.14 ± 0.03	2.2 ± 0.1	98 ± 14	19.0 ± 1.8	108 ± 9	67 ± 4
	k_{cat}/K_M (mM^{-1} min^{-1})	0.066 ± 0.037	7.0 ± 1.2	74 ± 31	166 ± 52	150 ± 37	105 ± 16
H_2O_2	K_M (mM)	3.21 ± 0.12	3.0 ± 0.4	7.9 ± 1.9	2.4 ± 0.5	0.43 ± 0.08	7.4 ± 1.5
	k_{cat} (min^{-1})	0.13 ± 0.01	1.9 ± 0.1	78 ± 8.9	15.2 ± 1.3	159 ± 12	71 ± 8
	k_{cat}/K_M (mM^{-1} min^{-1})	0.040 ± 0.004	0.63 ± 0.13	9.9 ± 3.6	6.4 ± 2.1	369 ± 100	9.6 ± 3.2

Data for ARS and PNS were previously reported and are added here to allow a full comparison [22,23] (*n* = 5).

To test the hypothesis of a one-electron abstraction mechanism as the first step of the bleaching process, the influence of several well-known redox mediators was also studied: 4-hydroxy-2,2,6,6-tetramethylpiperidine-1-oxyl (OH- TEMPO), *N*-hydroxyphthalimide (NPH), *N*-hydroxysuccinimide (NHS), and *N*-hydroxybenzotriazole (NHT). The results are summarized in Table 4. Only in the case of MG was an increase (three-fold) of the catalytic activity observed; in all the other cases, the redox-mediators mainly decreased IPS/MnTSPP catalysis (this quenching could be due to the lower reactivity of the oxygen-centered radical form of these mediators). Even in the best case, however, because the increase in catalytic activity was moderate, the industrial benefit of these redox mediators may not justify their environmental impact.

Table 4 Effect Of Well-Known Redox-Mediators On The Bleaching Of The Six Tested Dyes By Ips/Mntspp

Redox mediator	% Catalytic activity					
	ARS	PNS	XO	MB	MG	MO
OH-TEMPO	98	45	277	56	336	16
NHT	142	46	162	38	299	121
NHS	141	89	146	90	313	122
NPH	103	32	74	56	340	57

Results are expressed by referring to a 100% reference activity in the absence of any mediator. Data for ARS and PNS were previously reported and are added here to allow a full comparison [22,23].

The catalyst was also investigated for its reuse potential. The results are reported in Table 5. On average, a rapid decrease of catalytic activity was observed in the first 2–3 cycles. Subsequently, activity was stable at approximately 25-30% of maximum for at least 8–10 cycles. A potential explanation for this pattern is the possibility that a fraction of some dyes remains tightly adsorbed on the catalyst even when the supernatant is completely decolorized. This adsorbed dye would likely reduce the catalytic performance by steric hindrance and/or electrostatic repulsion of incoming substrate molecules. Nevertheless, the catalyst keeps a significant part of its catalytic activity during multi-cycle employment. This is an important quality for large-scale applications.

Table 5 Multi-Cycle Activity Of Ips/Mntspp

Catalytic cycle	% Catalytic activity					
	ARS	PNS	XO	MB	MG	MO
1	100	100	100	100	100	100
2	68	73	100	99	98	44
3	49	70	93	73	97	37
4	33	61	88	49	97	34
5	30	51	47	37	97	28
6	28	49	41	34	96	28
7	28	44	31	34	96	28
8	26	41	30	26	94	27
9	25	39	29	25	94	20
10	25	35	27	24	90	19

Catalytic activity of the IPS/MnTSPP adduct upon multicyclic use during the decolorization of the studied dyes ($n = 3$). Data for ARS and PNS were previously reported and are added here to allow a full comparison [22,23].

Significant removal of total organic carbon (TOC) was also observed during the bleaching by H_2O_2/IPS/MnTSPP, as shown in Figure 3. In the case of ARS, PNS, and MB, over 50% of TOC removal was observed in 24 hours, whereas lower decreases were recorded for XO and MG. Conversely, in the case of MO no TOC modification occurred during the catalysis. These results suggest that a decarboxylative degradation occurred for some of the dyes, whose TOC values decreased significantly during the bleaching experiments.

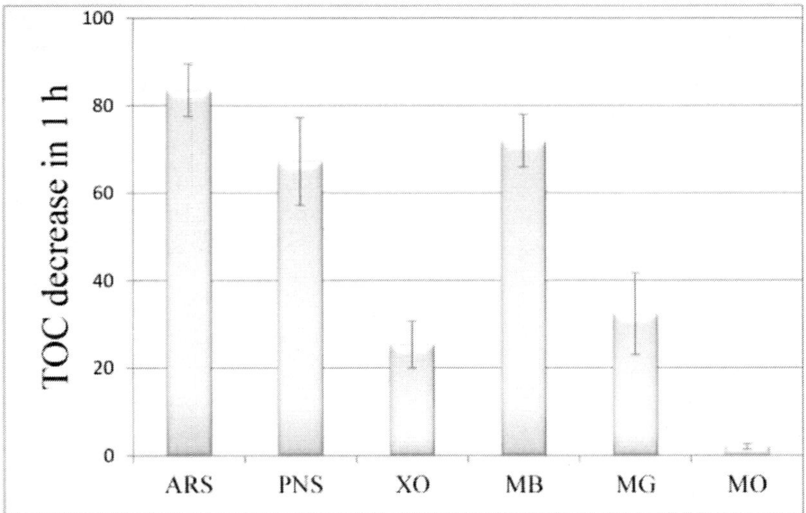

Figure 3 Total Organic Carbon (Toc) Decrease During The Bleaching Of The Textile Dyes By Ips/Mntspp (***n*** = 3).

IPS/MnTSPP was also tested for its ability to decolorize a complex mixture of all the six dyes, emulating realistic textile plant waste. The six dyes were each present in the saturating concentrations used above, leading to high impacting TOC (approximately 1580 mg/L). The changes in the UV–vis spectra (Figure 4) show a significant decolorization, although with a slower kinetics (approximately 5 hours for a 50% decrease of the main visible peak). However, the feasibility of using this treatment in an industrial process was confirmed by a significant TOC decrease (approximately 20%) in the same time frame.

2.3 Comparison with enzymatic bleaching

A full comparison with the enzymatic bleaching of the dyes was also performed. The ligninolytic enzymes tested were laccase (LC), lignin peroxidase, and the non-ligninolytic enzyme horseradish peroxidase (HRP). In addition, Mn(III) was also used as oxidizing agent, mimicking manganese peroxidase action [27]. These biological catalysts suffered from significant substrate inhibition, preventing their use at the dye concentrations used for IPS/MnTSPP. Usually, concentrations of dyes at least of one order of magnitude lower were necessary. These enzymes also suffer from H_2O_2 inhibition, needing lower concentrations of oxidizing substrate. Accordingly, under the same conditions, it is not surprising that enzymatic bleaching efficiency was significantly lower than IPS/MnTSPP bleaching efficiency (<1% dye removal in all cases). Enzymatic bleaching activity, under the conditions studied, is reported in Table 6, allowing a qualitative analysis of substrate specificity only.

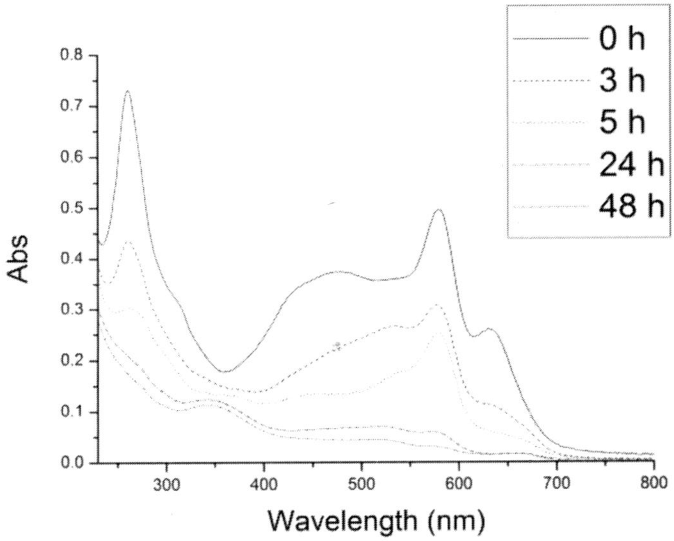

Figure 4 Changes In Uv–Vis Spectra During The Bleaching Of A Mix Of The Six Studied Dyes By Ips/Mntspp.

None of the examined enzymes were able to bleach all the studied dyes. For instance, PNS is resistant to LC, HRP, and Mn^{3+}, whereas only a partial degradation with concomitant irreversible enzyme inactivation is observed with LiP. MB underwent only partial oxidation with the enzymatic catalysts.

IPS/MnTSPP is able to effectively bleach all the dyes studied, as described above. This feature, combined with its ability to oxidize relatively high concentration of dye (in the millimolar range), and with its partial multi-cycling, may make IPS/MnTSPP catalysis a feasible alternative for industrial treatment of textile wastewaters.

Table 6 Enzymatic Bleaching Of The Dyes

Dye	LC	LC + H_2O_2	HRP	LiP	Mn^{3+}
ARS	+	++	+/−	+/−	+
PNS	-	-	-	+/−	-
XO	+	+	+	-	+
MB	-	+/−	+/−	+/−	-
MG	+	+	++	+/−	++
MO	+	+/−	+	+/−	+

In the conditions tested, ligninolytic enzymes present a narrower substrate specificity towards the dyes included in this study than IPS/MnTSPP. Legend: LC laccase, HRP horseradish peroxidase, LiP lignin peroxidase, ++ good substrate (complete bleaching < 1 h), + substrate, +/− weak substrate (bleaching < 10% in 24 h), - no substrate.

3. EXPERIMENTAL SECTION

3.1 Chemicals and instrumentation

All reagents were the best commercial grade available and were used without further purification. In particular, silica gel 100 came from Fluka (cat. No. 60746), and MnTSPP and FeTFPP were purchased from Sigma-Aldrich (cat. No. 441813 and 252913, respectively), PVA from Aldrich (cat. No. 363138), glutaraldehyde from Fluka (cat. no. 49629), horseradish peroxidase (E.C. 1.11.1.7) and lignin peroxidase (E.C. 1.11.1.14) were from Sigma-Aldrich (cat No. P-6782 and 42603, respectively). Laccase from *Pleurotus sajor-caju* was purified through two chromatographic steps (details in [28, 29]).

Spectrophotometric measurements were carried out with an UltroSpec 2100pro (Amersham Bioscience). Concentrations of the substrates were also measured by UV-HPLC and GC-MS, using previously described methods [30, 31].

3.2 Preparation of IPS/MnTSPP

The complex IPS/MnTSPP was synthesized as previously described [24]. Briefly, 10 g plain silica gel was functionalized by overnight reaction at 80°C with 10 mmol 3-(1-imidazolyl)propylcarbamoyl-3'-aminopropyl-triethoxysilane solubilized in 10 mL dioxane, in the presence of 10 μL *N*-methylmorpholine acting as the catalyst. The activated silica (IPS) was washed consecutively with 0.5 M HCl, H_2O, 0.1 M NaOH and H_2O again. The wet silica was then carefully dried overnight in a vacuum oven.

The bound MnTSPP was quantified by the difference in absorption at 400 nm (the λ_{max} of the MnTSPP Soret band), as described in [24].

3.3 Preparation of PP-PVA/FeTFPP

The complex PP-PVA/FeTFPP was synthesized as previously described [25]. Briefly, 1 g of aminopropyl crosslinked PVA (prepared as in [32]) was suspended in an excess of 0.05 M sodium acetate buffer pH 5, and treated with 0.1 mL 4-pyridinecarboxaldehyde and 0.5 g sodium cyanoborohydride. After 24 h of incubation at 25°C, the support was exhaustively washed with water, and dried at 50°C, to obtain PP-PVA. One-gram aliquots of PP-PVA were treated with 20 mg of FeTFPP, solubilized in 10 mL DMSO. The slurries were kept stirring in the dark for 24 h, washed exhaustively with DMSO, and carefully dried at 50°C.

Bound metalloporphine was quantified by difference through spectrophotometric measurement at 411 nm (ε_{411} = 115,000 M^{-1} cm^{-1} in DMSO) [25].

3.4 Activity measurements

Routine measurements of the catalytic activity of immobilized metalloporphines were performed through spectrophotometric assays. The suspension contained 5 mg of catalyst in 1 mL of 25 mM buffer solution containing 8.8 mM hydrogen peroxide and a proper concentration of the selected dye. The routine analyses were performed using the following concentrations of dyes (chosen to be only slightly higher than respective K_M, to better observe influences on catalytic activity): 0.29 mM ARS, 0.31 mM PNS, 2 mM XO, 0.15 mM MB, 1.5 mM MG, and 1.25 mM MO. During the kinetic experiments, wide ranges of substrate concentrations were studied (0.01-2 mM for the dyes, 0.1-40 mM for hydrogen peroxide).

The reaction mixture was then stirred at 25°C in the dark, and the reaction was monitored photometrically using the λ_{max} and molar extinction coefficients, reported in Table 1.

The actual hydrogen peroxide concentration was determined by $KMnO_4$ titration.

When the reaction pH was not 7, the pH was corrected to 7 with 0.1 mL of potassium phosphate buffer 1 M (pH 7) before an absorbance measurement. UV–vis spectra in the range 230–700 nm were also recorded. Different McIlvaine buffers were used in the pH range 3–8.

When used, redox mediators (4-hydroxy-2,2,6,6-tetramethylpiperidine-1-oxyl OH- TEMPO; *N*-hydroxyphthalimide NPH; *N*-hydroxysuccinimide NHS; and *N*-hydroxybenzotriazole NHT) were added to the reaction medium at a final concentration of 1 mM. For the MnP-like assays, all the experiments were repeated as above in the presence of 1 mM $MnSO_4$ and 50 mM malonic acid. During multi-cyclic runs, the catalysts were regenerated between consecutive cycles by repeated washings with water and 2-propanol and by drying at 50°C.

Total organic carbon determination was performed by automatic analyzer TOC 5000A (Shimadzu), equipped with Pt-Al_2O_3 oxidizing catalyst. The volume of injection was 25 μL. The calibration curve was obtained using a range of standard solutions of potassium phthalate (0–200 ppm).

Michaelis-Menten kinetic parameters were calculated by using R 2.5.1 software (*R Foundation for Statistical Computing*, Vienna).

3.5 Enzymatic assays

When horseradish peroxidase (HRP) was used, up to 20 E.U. were present in a final volume of 1 mL of 25 mM buffer (pH 4, 5, 6, or 7) with 8.8 mM H_2O_2. Due to significant substrate inhibition, lower dyes concentrations were used: 119 μM ARS; 31 μM PNS; 200 μM XO; 15 μM MB; 150 μM MG; 65 μM MO.

The same conditions were adopted for the experiments with laccase (23 E.U. added); every experiment was performed in the presence and absence of H_2O_2.

In the case of lignin peroxidase (LiP), 0.2 E.U. were present in the final volume of 1 mL, with 0.176 mM H_2O_2, due to strong substrate inhibition.

Moreover, to mimic manganese peroxidase (MnP) action, other experiments were conducted using Mn(III) as the putative oxidizing agent. One mM manganese triacetate was dissolved in 50 mM sodium malonate buffer, pH 6.5 [27], and the final mixture contained this solution along with dyes at the same concentrations, as mentioned above.

4. CONCLUSION

IPS/MnTSPP has proved able to efficiently degrade all tested dyes, regardless of their chemical class, under very mild conditions: standard pressure, room temperature and neutral pH, using no organic solvents, and the most environmental-friendly oxidant. The yields and time requirements of the process were quite promising for a potential industrial application. Furthermore, a multi-cycle capability was described, even if only 30% of catalytic activity is retained for 8–10 cycles.

In comparison to enzymatic systems, the biomimetic catalyst was more efficient and had wider substrate specificity. Therefore, after proper optimization of the process, this catalyst is worth consideration as a possible medium/large scale application in textile wastewater treatment.

ACKNOWLEDGMENTS

This research was partly funded by Fondazione Banco di Sardegna (Project Code 861/2012) and Regione Autonoma Sardegna (PO Sardegna FSE 2007–2013, L.R.7/2007, Bando Giovani Ricercatori, Project CRP1_27, and PO Sardegna FSE 2007–2013, L.R.7/2007, Ricerca di Base Bando 2008, Project CRP1_408).

COMPETING INTERESTS

The authors declare that they have no competing interests.

AUTHORS' CONTRIBUTIONS

PZ performed the main part of the experiments and drafted the manuscript. MP performed TOC and enzymatic assays. ACR, and AR helped analyzing the data and draft the manuscript. ES conceived the study, coordinated the study, and worked on data analysis and interpretation. All the authors read and approved the final manuscript.

REFERENCES

1 Forgacs E, Cserhati T, Oros G: Removal Of Synthetic Dyes From Wastewaters: A Review. *Environ Int* 2004, 30:953–971.

2 Azizi A, Moghaddam Mra, Arami M: Application Of Wood Waste For Removal Of Reactive Blue 19 From Aqueous Solutions: Optimization Through Response Surface Methodology. *Environ Eng Manag J* 2012, 11:795–804.

3 Brînzilă Ci, Ciobanu R, Brillas E: Effect Of Experimental Parameters On Crystal Violet Mineralization By Electro-Fenton Processes. *Environ Eng Manag J* 2012, 11:517–523.

4 Tsuda S, Murakami M, Matsusaka N, Kano K, Taniguchi K, Sasaki Yf: Dna Damage Induced By Red Food Dyes Orally Administered To Pregnant And Male Mice. *Toxicol Sci* 2001, 61:92–99.

5 Walthall Wk, Stark Jd: The Acute And Chronic Toxicity Of Two Xanthene Dyes, Fluorescein Sodium Salt And Phloxine B, To Daphnia Pulex. *Environ Pollut* 1999, 104:207–215.

6 Pierce J: Colour In Textile Effluents - The Origins Of The Problem. *J Soc Dye Colour* 1994, 110:131–133.

7 Colar La, Ilinoiu Ec, Manea F, Pode R, Lazau C: Photocatalytic Activity Of A Silver Doped Tio 2 Modified Zeolite In The Degradation Of Reactive Yellow 125 Azo Dye. *Environ Eng Manag J* 2012, 11:61–68.

8 Zucca P, Rescigno A, Olianas A, Maccioni S, Sollai F, Sanjust E: Induction, Purification, And Characterization Of A Laccase Isozyme From Pleurotus Sajor-Caju And The Potential In Decolorization Of Textile Dyes. *J Mol Catal B Enzym* 2011, 68:216–222.

9 Pakshirajan K, Jaiswal S, Das Rk: Biodecolourization Of Azo Dyes Using Phanerochaete Chrysosporium: Effect Of Culture Conditions And Enzyme Activities. *J Sci Ind Res* 2011, 70:987–991.

10 Córdoba A, Magario I, Ferreira Ml: Experimental Design And Mm2-Pm6 Molecular Modelling Of Hematin As A Peroxidase-Like Catalyst In Alizarin Red S Degradation. *J Mol Catal A Chem* 2012, 355:44–60.

11 Zhu L, Ghosh T, Park Cy, Meng Zd, Oh Wc: Enhanced Sonocatalytic Degradation Of Rhodamine B By Graphene-Tio 2 Composites Synthesized By An Ultrasonic-Assisted Method. *Chin J Catal* 2012, 33:1276–1283.

12 Oh Wc, Chen M, Cho K, Kim C, Meng Z, Zhu L: Synthesis Of Graphene-Cdse Composite By A Simple Hydrothermal Method And Its Photocatalytic Degradation Of Organic Dyes. *Chin J Catal* 2011, 32:1577–1583.

13 Asgher M, Kamal S, Iqbal Hmn: Improvement Of Catalytic Efficiency, Thermo-Stability And Dye Decolorization Capability Of Pleurotus Ostreatus Ibl-02 Laccase By Hydrophobic Sol Gel Entrapment. *Chem Centr J* 2012., 110:

14 Stengl V, Matys Grygar T, Henych J, Kormunda M: Hydrogen Peroxide Route To Sn-Doped Titania Photocatalysts. *Chem Centr J* 2012., 113:

15 Velmurugan R, Sreedhar B, Swaminathan M: Nanostructured Agbr Loaded Tio 2: An Efficient Sunlight Active Photocatalyst For Degradation Of Reactive Red 120. *Chem Centr J* 2011., 5:

16 Chiavola A: Textiles. *Water Environ Res* 2012, 84:1511–1532.

17 Zucca P, Rescigno A, Sanjust E: Ligninolytic Peroxidase-Like Activity Of A Synthetic Metalloporphine Immobilized Onto Mercapto-Grafted Crosslinked Pva Inspired By The Active Site Of Cytochrome P450. *Chin J Catal* 2011, 32:1663–1666.

18 Faria Al, Mac Leod Toc, Barros Vp, Assis Md: Hydrocarbon Oxidation Catalyzed By Iron And Manganese Porphyrins Anchored On Aminofunctionalized Supports. *J Brazil Chem Soc* 2009, 20:895–906.

19 Dos Santos Js, Faria Al, Da Silva Amorin Pm, La Luna Fm, Caiado Kl, Silva Doc E, Sartoratto Ppc, Assis Md: Iron(Iii) Porphyrin Covalently Supported Onto Magnetic Amino-Functionalized Nanospheres As Catalyst For Hydrocarbon And Herbicide Oxidations. *J Brazil Chem Soc* 2012, 23:1411–1420.

20 Díaz-Díaz G, Blanco-López Mc, Lobo-Castañón Mj, Miranda-Ordieres Aj, Tuñón-Blanco P: Hemo-Acrylic Polymers As Catalyst For The Oxidative Dehalogenation Of 2,4,6-Trichlorophenol. Chloroperoxidase's Mimic Imprinting Effects. *J Mol Catal A Chem* 2012, 353–354:117–121.

21 Tian P, Gao B, Chen Y: Synchronous Synthesis And Immobilization Of Pyridyl Porphyrins On Crosslinked Polystyrene Microspheres And Catalytic Oxidation Performance Of Immobilized Cationic Co-Porphyrins. *Chin J Catal* 2011, 32:483–489.

22 Zucca P, Vinci C, Sollai F, Rescigno A, Sanjust E: Degradation Of Alizarin Red S Under Mild Experimental Conditions By Immobilized 5,10,15,20-Tetrakis(4-Sulfonatophenyl)Porphine-Mn(Iii) As A Biomimetic Peroxidase-Like Catalyst. *J Mol Catal A Chem* 2008,288:97–102.

23 Zucca P, Vinci C, Rescigno A, Dumitriu E, Sanjust E: Is The Bleaching Of Phenosafranine By Hydrogen Peroxide Oxidation Catalyzed By

Silica-Supported 5,10,15,20-Tetrakis-(Sulfonatophenyl)Porphine-Mn(Iii) Really Biomimetic? *J Mol Catal A Chem* 2010, 321:27–33.

24 Zucca P, Mocci G, Rescigno A, Sanjust E: 5,10,15,20-Tetrakis(4-Sulfonato-Phenyl)Porphine-Mn(Iii) Immobilized On Imidazole Activacted Silica As A Novel Lignin-Peroxidase-Like Biomimetic Catalyst. *J Mol Catal A Chem* 2007, 278:220–227.

25 Zucca P, Sollai F, Garau A, Rescigno A, Sanjust E: Fe(Iii)-5,10,15,20-Tetrakis(Pentafluorophenyl)Porphine Supported On Pyridyl-Functionalized, Crosslinked Poly(Vinylalcohol) As A Biomimetic Versatile-Peroxidase-Like Catalyst. *J Mol Catal A Chem* 2009,306:89–96.

26 Cojocariu Am, Mutin Ph, Dumitriu E, Vioux A, Fajula F, Hulea V: Removal Of Dimethylsulfoxide From Wastewater Using Mild Oxidation With H2o2 Over Ti-Based Catalysts. *Chemosphere* 2009, 77:1065–1068.

27 Wariishi H, Valli K, Gold Mh: Manganese(Ii) Oxidation By Manganese Peroxidase From The Basidiomycete Phanerochaete-Chrysosporium - Kinetic Mechanism And Role Of Chelators. *J Biol Chem* 1992, 267:23688–23695.

28 Rescigno A, Zucca P, Flurkey A, Inlow J, Flurkey Wh: Identification And Discrimination Between Some Contaminant Enzyme Activities In Commercial Preparations Of Mushroom Tyrosinase. *Enzyme Microb Technol* 2007, 41:620–627.

29 Sollai F, Zucca P, Sanjust E, Steri D, Rescigno A: Umbelliferone And Esculetin: Inhibitors Or Substrates For Polyphenol Oxidases? *Biol Pharm Bull* 2008, 31:2187–2193.

30 Zucca P, Littarru M, Rescigno A, Sanjust E: Cofactor Recycling For Selective Enzymatic Biotrasformation Of Cinnamaldehyde To Cinnamyl Alcohol. *Biosci Biotechnol Biochem* 2009, 73:1224–1226.

31 Rescigno A, Casañola-Martin Gm, Sanjust E, Zucca P, Marrero-Ponce Y: Vanilloid Derivatives As Tyrosinase Inhibitors Driven By Virtual Screening-Based Qsar Models. *Drug Test Anal* 2011, 3:176–181.

32 Sanjust E, Cocco D, Curreli N, Rescigno A, Sollai F, Bannister Jv: Flavin-Grafted Poly(Vinyl Alcohol): Preparation And Properties. *J Appl Polym Sci* 2002, 85:2471–2477.

CHAPTER 10

Physichochemical and Low Stress Mechanical Properties of Silk Fabrics Degummed by Enzymes

Styliani Kalantzi[1], Diomi Mamma[1] and Dimitris Kekos[1]

[1] Biotechnology Technology, School of Chemical Engineering, National Technical University of Athens, Zografou Campus, Athens, Greece

1. INTRODUCTION

1.1. The Silk Road

Sericulture or silk production has a long and colorful history unknown to most people. For centuries the West knew very little about silk and the people who made it. For more than two thousand years the Chinese kept the secret of silk altogether to themselves. It was the most zealously guarded secret in history. According to Chinese tradition, the history of silk begins in the 27th century BCE. Its use was confined to China until the Silk Road opened at some point during the latter half of the first millennium BCE.

The writings of Confucius and Chinese tradition recount that in the 27th century BCE a silk worm's cocoon fell into the tea cup of the empress Leizu. Wishing to extract it from her drink, the young girl of fourteen began to unroll the thread of the cocoon. She then had the idea to weave it. Having observed the life of the silk worm on the recommendation of her husband, the Yellow Emperor, she began to instruct her entourage the art of raising silk worms, sericulture. From this point on, the girl became the goddess of silk in Chinese mythology [1].

Though silk was exported to foreign countries in great amounts, sericulture remained a secret that the Chinese guarded carefully. Consequently, other peoples invented wildly varying accounts of the source of the incredible fabric. In classical antiquity, most Romans, great admirers of the cloth, were convinced that the Chinese took the fabric from tree leaves. This belief was affirmed by Seneca the Younger in his Phaedra and by Virgil in his Georgics. Notably, Pliny the Elder knew better. Speaking of the bombyx or silk moth, he wrote in his Natural History "They weave webs, like spiders, that become a luxurious clothing material for women, called silk".

The earliest evidence of silk was found at the sites of Yangshao culture in Xia County, Shanxi, where a silk cocoon was found cut in half by a sharp knife, dating back to between 4000 and 3000 BCE. The species was identified as bombyx mori, the domesticated silkworm. Fragments of primitive loom can also be seen from the sites of Hemudu culture in Yuyao, Zhejiang, dated to about 4000 BCE. Scraps of silk were found in a Liangzhu culture site at Qianshanyang in Huzhou, Zhejiang, dating back to 2700 BCE. Other fragments have been recovered from royal tombs in the Shang Dynasty (c. 1600 - c. 1046 BCE).

During the later epoch, the Chinese lost their secret to the Koreans, the Japanese, and later the Indians, as they discovered how to make silk. Allusions to the fabric in the Old Testament show that it was known in western Asia in biblical times. Scholars believe that starting in the 2nd century BCE the Chinese established a commercial network aimed at exporting silk to the West. Silk was used, for example, by the Persian court and its king, Darius III, when Alexander the Great conquered the empire. Even though silk spread rapidly across Eurasia, with the possible exception of Japan its production remained exclusively Chinese for three millennia.

According to FAO estimates, the world raw silk production for the year 2010 was 164971 tonnes [2]. Approximately 98% of the world's production is in Asia and especially in Eastern Asia (Figure 1). China is the leader in raw silk production followed by India (Table 1).

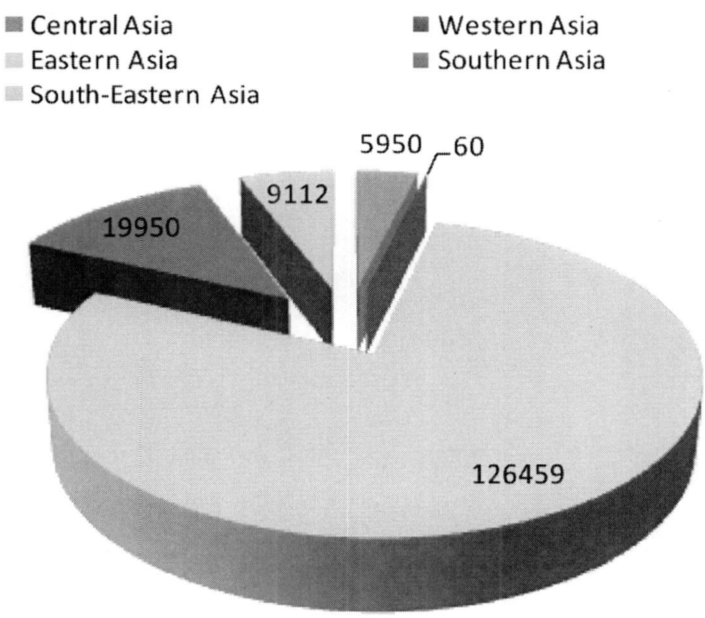

Figure 1. Production of raw silk across Asia (FAO, 2010)

Table 1. Raw silk producing countries [2].

Country	Production (tonnes)	Country	Production (tonnes)
China	126001	Japan	105
India	19000	Afghanistan	50

Country	Production (tonnes)	Country	Production (tonnes)
Viet Nam	7367	Kyrgyzstan	50
Turkmenistan	4500	Turkey	50
Thailand	1600	Cambodia	25
Brazil	1300	Italy	12
Uzbekistan	1200	Lebanon	10
Iran (Islamic Republic of)	900	Bulgaria	5
Democratic People's Republic of Korea	350	Greece	5
Tajikistan	200	Egypt	3
Indonesia	120		

1.2. Types Of Silk

Silk-producing insects have been classified on the basis of morphological clues, such as follicular imprints on the chorine egg, arrangement of tubercular setae on the larvae, and karyotyping data [3-4]. Classification based on phenotypic attributes is sometimes misleading because morphological features may vary with the environment [5]. Molecular marker-based analysis has been developed to distinguish genetic diversity among silkworm species [6-9]. Most commercially exploited silk moths belong to either the family Bombycidae or Saturniidae, in the order lepidoptera. Silkworms can be divided in three groups: (a) univoltine breed (one generation per year) which is usually found in Europe where due to the cold climate the eggs are dormant in winter and they are hatched in spring, (b) bivoltine breed (two generations per year) usually found in Japan China and Korea, where the climate is suitable for developing two life cycle per year and (c) multivoltine breed (up to eight generations per year) usually found in tropical zone.

The finest quality raw silk and the highest fiber production are obtained from the commonly domesticated silkworm, *Bombyx mori*, which feeds on the leaves of the mulberry plant, *Morus* spp. Other than the domesticated *B. mori*, silk fiber production is reported from the wild non-mulberry saturniid variety of silkworms. Saturniid silks are of three types: tasar, muga, and eri (Table 2).

Table 2. Commercially exploited sericigenous insects of the world and their food plants [**10**].

Common Name	**Scientific Name**	**Origin**	**Primary Food Plant(s)**
Mulberry Silkworm	*Bombyx mori*	China	Morus indica, M. alba M.multicaulis, M.bombycis

Common Name	**Scientific Name**	**Origin**	**Primary Food Plant(s)**
Tropical Tasar Silkworm	*Antheraea mylitta*	India	Shorea robusta, Terminalia tomentosa T. arjuna
Oak Tasar Silkworm	*Antheraea proylei*	India	Quercus incana, Q. serrate Q. himalayana, Q. leuco tricophora Q. semicarpifolia, Q. grifithi
Oak Tasar Silkworm	*Antheraea frithi*	India	*Q. dealdata*
Oak Tasar Silkworm	*Antheraea compta*	India	*Q. dealdata*
Oak Tasar Silkworm	*Antheraea pernyi*	China	*Q. dendata*
Oak Tasar Silkworm	*Antheraea yamamai*	Japan	*Q. acutissima*
Muga Silkworm	*Antheraea assama*	India	Litsea polyantha, L. citrate Machilus bombycine

Common Name	Scientific Name	Origin	Primary Food Plant(s)
Eri Silkworm	*Philosamia ricini*	India	Ricinus communis, Manihot utilisma Evodia fragrance

The tasar silkworms are of two categories-Indian tropical tasar, *Antheraea mylitta*, which feeds on the leaves of *Terminalia arjuna*, *Terminalia tomantosa*, and *Shorea robusta*, and the Chinese temperate oak tasar, *Antheraea pernyi*, which feeds on the leaves of *Quercus* spp. and *Philosamia* spp. Indian tropical tasar (Tussah) is copperish colour, coarse silk mainly used for furnishings and interiors. It is less lustrous than mulberry silk, but has its own feel and appeal. Oak tasar is a finer variety of tasar silk [11].

Muga silk is produced by the multivoltine silkworm, *Antheraea assamensis* (also called *A. assama*), which feeds mainly on *Machilus* spp (Table 2). Muga is a golden yellow colour silk. Muga culture is specific to the state of Assam (India) and an integral part of the tradition and culture of that state. The muga silk, a high value product is used in products like sarees, mekhalas, chaddars, etc. [10]. Eri silk is produced by *Philosamia* spp. (*Samia* spp.), whose primary host plant is the castor (*Ricinus* spp.) (Table 2) [11]. The luster and regularity of *B. mori* silk makes it superior to the silk produced by the non mulberry saturniid silkworms, although non-mulberry silk fibers are also used commercially due to their higher tensile strength and larger cocoon sizes. Spider also produced silk fibers that are strong and fine, but have not been utilized in the textile industries [11].

1.3. Structure Of The Silk Fibre

The cocoons of the mulberry silkworm *B. mori* are composed of two major types of proteins: fibroins and sericin. Fibroin, the 'core' protein constitutes over 70% of the cocoon and is a hydrophobic glycoprotein [12] secreted from the posterior part of the silk gland (PSG) [13].

The fibroin, rich in glycine (43.7%), alanine (28.8%) and serine (11.9%), is composed of a heavy chain (~325 kDa), a light chain (~25 kDa) and a glycoprotein, P25, with molar ration of 6:6:1. The heavy and light chains are linked by a disulfide bond. P25 associates with disulfide-linked heavy and light chains primarily by non-covalently hydrophobic interactions, and plays an important role in maintaining integrity of the complex [14]. The light chain has a non-repetitive sequence and plays only a marginal role in the fiber. The heavy chain contains very long stretches of Gly-X repeats (with residue X being Ala in 64%, Ser in 22%, Tyr in 10%, Val in 3%, and Thr in 1.3%) that consist of 12 repetitive domains (R01–R12) separated by short linkers. It is an antiparallel, hydrogen bonded β-sheet and yields the X-ray diffracting structure called the "crystalline" component of silk fibroin [15]. Silk is a typical representative of β-sheet. Each domain consists of sub-domain hexapeptides including: GAGAGS, GAGAGY, GAGAGA or GAGTGA (G is glycine, A is alanine, S is serine and Y is tyrosine) [16]. In contrast, the 151 residues of the N-terminal, 50 residues of the C-terminal, and the 42-43 residues separating the 12 domains are non-repetitive and "amorphous" [17]. Silk fibroin can exist as three structural morphologies termed silk I, II, and III where silk I is a water soluble form and silk II is an insoluble form consisting of extended β-sheets. The silk III structure is helical and is observed at the air-water interface. In the silk II form, the 12 repetitive domains form anti-parallel b-sheets stabilized by hydrogen bonding [18]. Due to the highly oriented and crystalline structure of Silk II, silk fibroin fiber is hydrophobic and has impressive mechanical properties. When controllably spun, its mechanical property may be nearly as impressive as spider dragline silk [19].

Sericins, the 'glue' proteins constitute 20–30% of the cocoon, and are hot water-soluble glycoproteins that hold the fibers (fibroin) together to form the environmentally stable fibroin–sericin composite cocoon structure [20-22]. Sericin, secreted in the mid-region of the silk gland, comprises different polypeptides ranging in weight from 24 to 400 kDa depending on gene coding and post-translational modifications and are characterized by unusually high serine content (40%) along with significant amounts of glycine (16%), [23-24]. Three major fractions of sericin have been isolated from the cocoon, with molecular weights 150, 250, and 400 kDa [24]. Sericin remains in a partially unfolded state, with 35% β-sheet and 63% random coil, and with no α-helical content [18].

The amino acid compositions of fibroin and sericin have been published, with somewhat differences from paper to paper for some specific amino acid contents [16, 25-26].

The chemical composition of raw silk obtained from the silk worm *Bombyx mori* is presented on Table 3. Silk is produced in several countries and the fibres from different regions contain different amounts of sericin which exhibits diverse chemical and physical properties [27].

Table 3. Composition of raw silk from the silk worm *Bombyx mori* [**28-29**].

Component	
Fibroin	70-80
Sericin	20-30
Wax	0.4-0.8
Carbohydrate	1.2-1.6
Inogranic matter	0.7
Pigments	0.2

1.4. Silk Degumming

Silk processing from cocoons to the finished clothing articles consists of a series of steps which include: reeling, weaving, degumming, dyeing or printing, and finishing. Degumming is a key process during which sericin is totally removed and silk fibres gain the typical shiny aspect, soft handle, and elegant drape highly appreciated by the consumers. In addition, the existence of sericin prevents the penetration of dye liquor and other solutions during wet processing of silk. Also, it is the main cause of adverse problems with biocompatibility and hypersensitivity to silk [17]. Furthermore, to prepare pure silk fibroin solution for silk-based biomaterials, separation of silk fibroin fiber from the sericin glue, is a critical step, since (a) residual sericin causes inflammatory responses and (b) non-degummed fibers are resistant to solubilization [30].

The industrial process takes advantage of the different chemical and physical properties of the two silk components, fibroin and sericin. While the former is water-insoluble owing to its highly oriented and crystalline fibrous structure, the latter is readily solubilized by boiling aqueous solutions containing soap [27], alkali [31], synthetic detergents [32]. However, the higher temperature (95°C) and an alkaline pH (8-9) in the presence of harsh chemicals in the treatment bath impose a markedly unnatural environment on the silk, and thus cause partial degradation of fibroin. Fibre degradation often appears as loss of aesthetic and physical properties, such as dull appearance, surface fibrillation, poor handle, drop of tensile strength, as well as uneven dyestuff absorption during subsequent dyeing and printing [33]. More importantly, the large consumption of water and energy contribute to environmental pollution. These costs have fueled interest in developing a new, effective degumming method which minimizes these adverse effects (Table 4).

Many researches have been performed on degumming and finishing of silk fiber using acid agents for enhancing the physical properties of silk [34-36]. It has been pointed out that the action of organic acids is generally milder and less aggressive than the action of alkali. Khan et al. [37] investigated silk degumming using citric acid. The surface morphology of silk fiber degummed with citric acid was very smooth and fine, showed perfect degumming (almost complete removal of sericin) like traditional soap-alkali method and the tensile strength of silk fiber was increased after degumming. Tartaric and succinic acids demonstrate efficient sericin removal while retaining the intrinsic properties of the fiber. Freddi et al. [34] studied on the degumming of silk fabrics with tartaric acid and showed the excellent performances of tartaric acid, both in terms of silk sericin removal efficiency and of intrinsic physico-mechanical characteristics of silk fibers. The degummed silk fabric with tartaric acid exhibited a good luster and a 'scroopier' handle in compared with soap degummed fabric. They also demonstrated that dyeability with acid dyes and comfort properties (such as wicking, wettability, water retention and permeability) are also enhanced and concluded that the acid degumming process shows potential for possible industrial application. However, there is a tendency toward a gradual decrease of tenacity and elongation values with increasing tartaric acid in the bath.

Table 4. Degumming methods of raw silk

Degumming method	**References**
Traditional (alkali, soap, synthetic detergents)	[27, 31-32]
Microwave irradiation	[42]
Plasma method	[41]
Organic acids	[34-37]
Ultrasound method	[43-44]
Enzymatic method	[33, 46-52]

Low-temperature plasma treatment has been well studied at research level in textile in the last years due to its rapid, water and chemical free process, as well as resource conservation, though it is not yet really used and well established in industry [38-40]. Long et al., [41] reported that degumming efficiency and properties of silk fabric after low-pressure argon plasma treatment were comparable to the conventional wet-chemical treatment process. Unfortunately, plasma methods result in a notable etching effect from physical bombardments and chemical reactions by excited plasma species on sericin layers.

Microwave irradiation and ultrasound are techniques that have been investigated for their performance as degumming agents by several researchers. Microwave treatment of silk resulted in increased weight loss followed by a decrease in strength of the filaments, whereas the elongation increases. This can be explained by the fact that sericin is acting as an adhesive and working as a coating and wrapping material around the fibroin [42].

Ultrasound has been widely used in chemistry and the dyeing, finishing, and cleaning industries because of its obvious advantages in particle treatment, including dispersion and agglomeration effects. Ultrasonic method combined also with natural soaps (olive oil, turpentine and daphne soaps) or proteolytic enzymes (alcalase and savinase) enables an effective clearance in the degumming process, it facilitates the removal of the substances existing on the raw silk like dirt and sericin and yields positive results in terms of weight loss, whiteness degree and mechanical properties [43-44].

The increasing awareness of legislators and citizens for the ecological sustainability of industrial processes has recently stimulated the interest of scientists and technologists for the application of biotechnology to textile processing [45]. In recent years, various studies have dealt with the removal of sericin by using proteolytic enzymes since they can operate under mild conditions and low temperatures which save energy in comparison to the traditional method. Enzymes act selectively and can attack only specific parts of sericin to cause proteolytic degradation. So the pattern of soluble sericin peptides obtained by degumming silk changes as a function of the kinds of enzyme used, attributing to the different target cleavage of the enzymes.

Several acidic, neutral, and alkaline proteases have been used on silk yarn as degumming agents. Alkaline proteases performed better than acidic and neutral ones in terms of complete and uniform sericin removal, retention of tensile properties, and improvement of surface smoothness, handle, and lustre of silk [46-48]. Enzyme degummed silk fabric displayed a higher degree of surface whiteness, but higher shear and bending rigidity, lower fullness, and softness of handle than soap and alkali degummed fabric, owing to residual sericin remaining at the cross over points between warp and weft yarns [49]. Freddi et al. [33] applied acidic, neutral, and alkaline proteases to silk degumming and found that alkaline and neutral proteases performed better than acidic proteases in terms of complete sericin removal. After complete sericin removal with proteolytic methods, the quality of appearance and retention of tensile properties is expected to be superior to those silks degummed through traditional methods due to less chemical and physical stress applied to the silk during enzymatic processing. Nakpathom et al., [50] degummed Thai *Bombyx mori* silk fibers with papain enzyme and alkaline/soap and reported that the former exhibited less tensile strength drop and gave higher color depth after natural lac dyeing, especially when degumming occurred at room temperature condition. Alcalase, savinase, (two commercial proteolytic preparations) and their mixtures also proved to be feasible for degumming applications [51]. Gulrajani et al., [52] degummed silk with the combination of protease and lipase enzymes, and obtained efficient de-waxing and degumming effects, while maintaining favorable wettability of silk fibers.

Silk degumming is a high resource consuming process as far as water and energy are concerned. Moreover, it is ecologically questionable for the high environmental impact of effluents. The development of an effective degumming process based on enzymes as active agents would entail savings in terms of water, energy, chemicals, and effluent treatment. This could be made possible by the milder treatment conditions, the recycling of processing water, the recovery of valuable by-products such as sericin peptides, and the lower environmental impact of effluents [33]. However, the limitations of higher cost of enzymes compared to chemicals and the necessary continuous use of enzymes may limit the development of industrial processes using proteolytic degumming methods [41,49].

1.5. Potential Applications Of Sericin

Sericin is at present an unutilized by-product of the textile industry and the discarded degumming wastewater also ultimately leads to environmental contamination due to the high oxygen demand for its degradation by microbes [53]. It is estimated that out of the 1 million tons (fresh weight) of cocoon production worldwide, or about 400 000 tons of dry cocoon, approximately 50 000 tons of sericin could be recovered from the waste solution [54]. If sericin was recovered, perhaps it could be used as a 'value added' product for many sericin-derived products and purposes [55] and this would also be beneficial in terms of the economy and the environment.

Limitations on devising specific applications are caused by its ability to exist in many forms that depend on its method of extraction and purification, etc. Each specific application requires a particular form so it will be necessary to devise and understand how to prepare consistent products suitable for each. Non-textile applications of sericin range from cosmetics to biomedical products, which include its use in anticancer drugs, anticoagulants, and cell culture additives, for its antioxidant properties [11,56]. Furthemore, its ability to form crosslink or blends with other polymers to produce more effective films that can be used for new drug delivery methods with reduced immunogenicity and increased drug stability or even new food packaging materials worth further investigation [56].

2. MATERIALS & METHODS

2.1. Silk Fabric

The treatments were carried out on a 100% raw silk fabric (crêpe). Construction parameters are listed on Table 5.

2.2. Enzymes

Esperase® 8.0L and Lipolase® Ultra 50T were kindly provided from Novo Nordisk Co. (Bagsvaerd, Denmark). Papain was purchased from Sigma. All other chemicals were laboratory-grade (analytical reagents, Sigma).

Table 5. Construction properties of silk fabric

	Wrap yarn	**Weft yarn**
Number of ends (cm^{-1})	46	110
Fabric weight ($g{\cdot}m^{-2}$)	89.27	

2.3. Enzyme Activities

The proteolytic activity was assayed spectrophotometrically with azocasein as a substrate [57]. One unit of activity was defined as the amount of enzyme required to produce a 0.1 increase in absorbance at 440 nm under the assay conditions (50°C and pH 8.0).

Lipase activity was determined against p-nitrophenyl- propionate (pNPP) at pH 7.0 and 25°C [58]. The release of p-nitrophenol was monitored spectrophotometrically at 410 nm with the aid of a microplate reader (Molecular Devices Corporation, Sunnuvale, USA). The reaction was initiated by adding 10 μL of properly diluted enzyme to 190 μL of substrate solution (0.4 mM). Control reactions with inactivated lipase were used to correct nonenzymatic pNPP hydrolysis. One unit of activity was defined as the amount of enzyme which released 1 μmol of product per minute under the conditions described.

2.4. Soap Degumming (Conventional Method)

Silk fabrics were soaked overnight in a solution of 5 $g{\cdot}l^{-1}$ Marseille soap at pH 9.5 at a liquor-to-fabric ratio 40:1. Next day the silk fabrics were degummed in a boiled alkaline solution containing 10 $g{\cdot}l^{-1}$Marseille soap and 1 $g{\cdot}l^{-1}$ sodium carbonate for 2 h at a liquor-to-fabric ratio 40:1 and pH 9.5. Degummed silk fabrics were first rinsed at 50°C with 1 $ml{\cdot}l^{-1}$ ammonia and consequently two times at 40°C with 1 $ml{\cdot}l^{-1}$ ammonia. Finally the fabrics were rinsed with cool water.

2.5. Enzymatic Degumming

A fabric sample of approximate weight 3.0 g was immersed in Tris-HCl buffer (50 mM) at pH 8.0. A non-ionic wetting agent (Sadopane SF 0.1% w/v) and the appropriate amount of enzyme(s) supplemented the solution. The liquor-to-fabric ratio was adjusted to 40:1, and the mixture was incubated at 50°C and 50 rpm. Blank samples were obtained by treating silk with buffer alone, without enzyme. Enzyme dosage and treatment time were changed. Inactivation of the enzyme(s) was carried out in hot distilled water for 10 min. At the end of the treatment, silk fabrics were rinsed with distilled water and dried at room temperature. All degumming tests were performed in duplicate.

2.6. Bleaching Of Degummed Silk Fabrics

The enzymatic and conventional degummed silk fabrics were bleached using: 0.4% (o.w.g) bleaching agent (Belphor BH); 0.5% (o.w.g) stabilizator (Sifa FL); 20 $ml{\cdot}l^{-1}$ H_2O_2; 3 $g{\cdot}l^{-1}$ Na_2CO_3; liquor-to-fabric ratio (30:1); wetting agent Clariant Sandoclean PC-FL (1 $g{\cdot}l^{-1}$).

The bleaching agent was mixed with the stabilizator, H_2O_2 and the Na_2CO_3. The mixture was stirred and heated at 40°C. The silk fabrics were soaked in the mixture and stirred until the temperature reached 90°C. The fabrics were left for 100 min. The fabrics were removed and rinsed initially with hot tap water and finally they were immersed in water with 2-3 drops of formic acid.

2.7. Weight Loss And Degumming Efficiency Determination

Fabric weight loss was recorded as dried sample weight loss. The drying conditions were 105°C in an air-circulated oven for 1 h. The samples were weighed, after cooling in a desiccator. The following equation (Eq. (1)) was used to calculate the weight loss (wt%):

$$Wt\% = (\frac{W1 - W2}{W1}) \times 100 \quad (1)$$

where, *W1* and *W2* are the weights of the fabric before and after treatment, respectively [59].

The efficiency of the degumming was calculated through a comparison of the enzyme process for silk fabrics with the standard method (degumming using Marseille soap), using the following equation (Eq. (2)):

$$DegumEff = \frac{W_E}{W_{MST}} \quad (2)$$

where, *DegumEff* is the Degumming Efficiency (%), *WE* is the percentage of weight loss by the enzyme treatment and *WMTS* is the percentage of weight loss by the Marseille soap treatment [51].

The degumming with Marseilles soap was taken as the standard 100% weight loss.

2.8. Wettability: Drop Test

Wettability of the fabric was measured, by means of the "drop test" before and after the degumming process. The dried samples at room temperature were tested using AATCC Test Method 39-1980 (evaluation of wettability) [60]. The time period (in sec) between the contact of the water drop with the fabric and the disappearance of the water drop into the fabric was counted as the wetting time. The time of drop disappearance was averaged from measurements in different points of the fabric sample. Wetting times equal or less than 1 sec were considered as indication of adequate absorbency of the fabrics [61]. All measurements were performed in triplicate.

2.9. Whiteness

The whiteness index (Berger degree) of the fabrics was determined using a reflectance measuring Datacolor apparatus at standard illuminant D65 (LAV/Spec. Excl.,d/8, D65/10o) [62].

2.10. Crystallinity Index (Cri)

An X-ray diffractometer (Siemens D5000), was used in order to determine the Crystallinity Index, using copper Ka radiation. The angles scanned were 10–30° at 0.01°/s. The Crystallinity Index was determined according to the empirical method of Segal et al. [63] applying Eq. (3)

$$CrI = \left(\frac{I_{002} - I_{am}}{I_{002}}\right) \times 100 \qquad (3)$$

where I_{002} is the peak intensity from the lattice plane, and I_{am} is the peak intensity of amorphous phases.

Triplicate sets of data were used to establish the relative error associated with the X-ray diffraction method [63].

2.11. The Kawabata Evaluation System

Basic mechanical properties namely tensile, bending and shearing were measured by the KES-FB system under high sensitivity conditions. The temperature was 20 ± 0.5°C, and relative humidity is 65% ± 5. The properties measured were shear rigidity (*G*, $gf{\cdot}cm{\cdot}degree^{-1}$), bending rigidity per unit length (*B*, $gf{\cdot}cm^{2}{\cdot}cm^{-1}$) and extensibility (*EMT*, %) at 500 $gf{\cdot}cm^{-1}$. All measurements were performed in triplicate.

3. RESULTS AND DISCUSSION

3.1. Characteristics Of The Enzymes Used For Degumming

The commercial enzyme preparations used for degumming of the silk fabric are listed in Table 6. Novozymes launched Lipolase® in 1988, the first commercial lipase developed for the detergent Industry. Lipolase® was the first lipase produced by recombinant DNA technology. This lipase, originating from *Thermomyces lanuginosus*, formerly *Humicola lanuginosa*, was expressed in*Aspergillus oryzae*. This enzyme is widely used in detergent formulations to remove fat-containing stains and it also has a broad range of substrate specificity. Furthermore is stable in proteolytic wash solutions. Novozymes launched two variants of Lipolase® issued from rational protein design: Lipolase® Ultra and LipoPrime™. These variants were also expressed in *A. oryzae* [64]. The enzyme preparation used in the present study (Lipolase® Ultra 50T) exhibited optimal pH and temperature of 9.0 and 50°C respectively. Furthermore, it was stable at temperatures 30-50°C and pH values of 7.0-10.0 (the remaining activity was measured after incubation for 24 h at the above mentioned temperatures and the pH values).

Esperase® 8.0L is also a product of Novozymes. It is a bacterial serine type alkaline protease produced by *Bacillus* sp. Esperase® is characterized by excellent perfomance at elevated temperature and pH. It exhibited optimal pH and temperature of 10.0 and 70°C respectively. The enzyme preparation was stable (>90% of its original activity) at pH values of 7.0 and 8.0 and temperatures of 30-50°C (the remaining activity was measured after incubation for 24 h at the above mentioned temperatures and the pH values).

Papain is a cysteine protease, isolated from Papaya (*Carica papaya*) Latex. Papain consists of a single polypeptide chain with three disulfide bridges and a sulfhydryl group necessary for activity of the enzyme. The pH optimum of papain was found 8.0 while temperature optimum at 50ºC. Papain was stable at pH values of 7.0 and 8.0 and temperatures of 30-50ºC (the remaining activity was measured after incubation for 24 h at the above mentioned temperatures and the pH values).

Table 6. Enzymes used for silk degumming

Enzyme	Origin	Characteristics	Activity[a]	pH[a]	T (oC)[a]
Papain	*Carica papaya*	Cysteine protease	1.15 $U \cdot mg^{-1}$ dry matter	7.0-9.0 (8.0)	30-50 (50)
Esperase® 8.0L	*Bacillus* sp.	Serine-type protease	19.3 $U \cdot ml^{-1}$	7.0-8.0 (10.0)	30-50 (70)
Lipolase® Ultra 50T	Aspergillus oryzae (genetically modified)	Lipase	6.1 $U \cdot ml^{-1}$	7.0-9.0 (9.0)	30-50 (50)

[a] Optimum pH and temperature values in parentheses.

3.2. Protease Treatment Of Silk Fabrics: Effect Of Enzyme Dosage

Raw silk fabrics were degummed with Papain and Esperase® 8.0L using different enzyme loadings (expressed as Units of protease per g of silk fabric, $U \cdot g^{-1}_{fabric}$) and the physicochemical (weight loss, crystallinity index, whiteness, whiteness after bleaching) and low-stress mechanical properties (Kawabata evaluation system) of the preterated fabrics were assessed. Untreated and conventionally degummed (using Marseille soap) silk fabrics were used as controls.

3.2.1. Weight Loss

The weight loss (or degumming loss) represents a quantitative evaluation of the degumming efficiency after standard or enzymatic degumming. The effect of enzyme dosage on the extent of sericin removal was studied by treating silk fabric samples for 60 min with different amounts of the two proteases. The results are depicted in Figure 2 both as weight loss (Fig. 2a) and degumming efficiency (Fig. 2b).

Degumming loss increased linearly as the amount of papain increased, attaining a value of 10.2 % (w/w) (degumming efficiency 50%). Esperase® 8.0L was more efficient in sericin removal as judged by the weight loss (Fig. 2a). Maximum value (20.1%, w/w, degumming efficiency nearly 99%) was obtained at enzyme loading of 75 $U{\cdot}g^{-1}{}_{fabric}$. Degumming loss of silk fabrics treated with Marseille soap attained a value of 20.4% (w/w).

Without enzymes, the degumming loss was negligible, owing to the low treatment temperature. In fact, it is well known that sericin can be removed by using water alone, but high temperature is needed to attain complete degumming (110-120°C, under pressure).

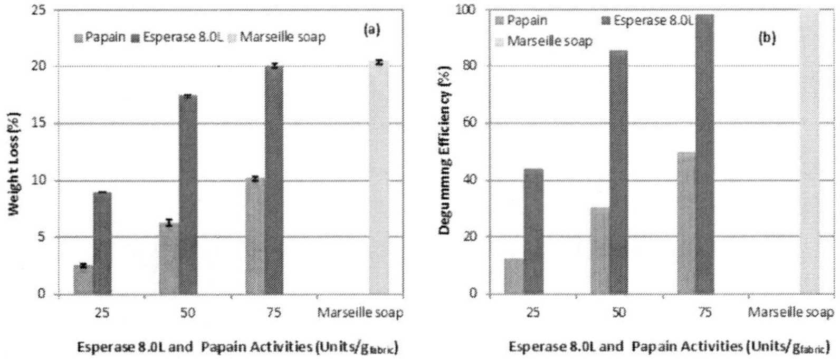

Figure 2. a) weight loss of silk fabrics treated with different types of proteases at enzyme dosages ranged from 25 to 74 $U{\cdot}g^{-1}{}_{fabric}$ and (b) degumming efficiency of the enzymatic process

The amount of sericin in raw silk measured in terms of weight loss varies between 17and 38%. Some of the good Chinese and Japanese varieties show about 17 to 17.5%, while yellow Italian silk has about 23% sericin and some Thai varieties have as high as 38% [27].

3.2.2. Wettability and Crystallinity

Wettability was a function of both enzyme dosage and type of protease. Silk fabrics treated with Esperase® 8.0L at 75 $U{\cdot}g^{-1}_{fabric}$ showed adequate water absorbency (<1 sec). On the other hand silk fabrics treated with papain showed higher wetting times namely more hydrophobic fabrics. Although the implementation of higher papain activities formed silk fabrics with lower wetting times (Table 7)these values were higher than 1 sec, which is the maximum wetting time required for efficient dyeing and finishing. The lowest wetting time achieved using papain was 3.65 sec at the highest papain activity (75 $U{\cdot}g^{-1}_{fabric}$).

Table 7. Wetting time and Crystallinity Index of silk fabrics treated conventionally (Marseille soap) or enzymatically.

Treatment	**Wetting time (sec)**	**Crystallinity Index (%)**
Papain ($U{\cdot}g^{-1}{}_{fabric}$)		
25	6.12 ± 0.25	61.73 ± 1.48
50	4.16 ± 0.16	62.27 ± 0.98
75	3.65 ± 0.19	63.90 ± 0.89
Esperase® 8.0L ($U{\cdot}g^{-1}{}_{fabric}$)		

Treatment	Wetting time (sec)	Crystallinity Index (%)
25	1.80 ± 0.20	63.90 ± 1.30
50	1.36 ± 0.15	66.61 ± 1.10
75	<1	66.72 ± 1.20
Marseille soap	<1	66.48 ± 1.35
No enzyme	6.85 ± 0.20	63.90 ± 1.27

The microstructure of cotton fabrics was investigated by X-ray diffraction. The results are presented on Table 7. The diffraction curves of all silk fibers exhibit the typical pattern of a silk II crystal with high crystallinity [19].

Silk fabrics treated with Esperase® 8.0L exhibited higher Crystallinity Index values compared to papain treated ones. Furthermore, at higher Esperase® 8.0L the crystallinity of the fabric was superior to that treated conventionally (Marseille soap).

3.2.3. Whiteness and Whiteness After Bleaching With H_2O_2

Natural coloring matters present in silk are associated mainly with sericin and hence are eliminated during degumming. The natural colouring matter of silk can be roughly divided into yellow, green and brown pigments. However the residual pigments are adsorbed by fibroin and hence silk fabrics made from yellow raw silk after degumming are not white but have a cream colour [65]. Lustre is one of the most important properties of silk, hence the method of degumming is significant.

A slight increase in whiteness was observed after treatment with the two different proteases compared to the fabric treated in the absence of enzyme. Esperase® 8.0L exhibited better results compared to papain. However, whiteness of silk fabrics treated with Marseille soap was superior to that of enzymatically treated fabrics (Figure 3a).

The results are in contrast to those reported by Chopra et al., [49] who demonstrated that enzyme treated samples rate marginally better than the soap-treated samples.

Bleaching of silk is for white and pastel shades only. Degummed silk fabrics present a slightly off-white in colour, because of some sericin, which is stubbornly stuck to the fibrin [65]. After enzymatic and conventional degumming the silk fabrics were bleached with H_2O_2. The results indicated that the whiteness after bleaching of all Esperase® 8.0L treated fabrics were the highest (Figure 3b).

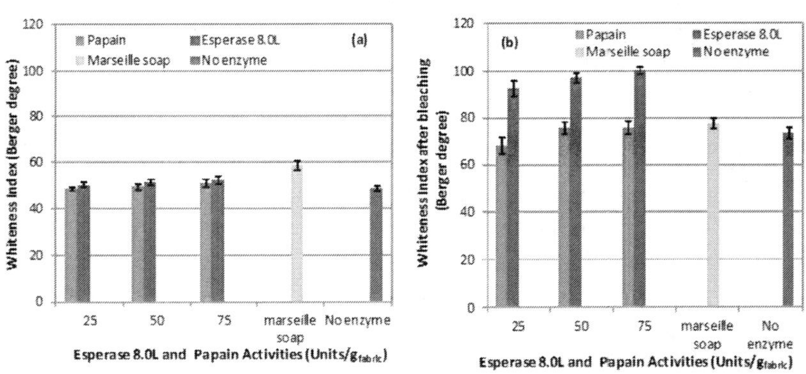

Figure 3. a) Whiteness Index and (b) Whiteness Index after bleaching with H_2O_2 of degummed silk fabrics using Marseille soap (conventional treatment) or papain and Esperase® 8.0L at different loadings.

3.2.4. Bending Property

Fabric bending property is apparently a function of the bending property of its constituent yarns. *B* is bending rigidity, a measure of a fabric's ability to resist bending deformation. In other words, it reflects the difficulty with which a fabric can be deformed by bending.This parameter is particularly critical in the tailoring of lightweight fabrics. The higher the bending rigidity, the higher a fabric's ability to resist when it is bent by external force like what happens during fabric manipulation in spreading and sewing.

Enzymatically treated silk fabrics exhibited higher *B* values compared to silk fabrics treated only with buffer solution (no enzyme) (Table 8). Bending rigidity was affected by both type of protease and enzyme concentration. Increasing enzyme concentration resulted in higher *B* values, namely in more stiff fabrics. Silk fabrics treated with papain exhibited 25-116% increase in *B* values while silk fabrics treated with Esperase® 8.0L showed higher increase namely 44-137% in comparison with the buffer treated silk fabrics. Enzymatically treated silk fabrics are less rigid, softer compared to conventionally degummed fabrics.

Chopra et al., [49] reported that enzyme treated silk fabrics exhibited increased bending rigidity compared to soap treated ones. In the present study, even though an increase in bending rigidity was observed through the different experimental conditions, fabric treated with Marseille soap exhibited the highest value of this property.

3.2.5. Shear Property

Shear rigidity *G* provides a measure of the resistance to rotational movement of the warp and weft threads within a fabric when subjected to low levels of shear deformation. The lower the value of *G*, the more readily the fabric will conform to three-dimensional curvatures.

Enzymatically treated silk fabrics showed lower *G* values compared to untreated and conventionally treated silk fabrics. Shear rigidity (*G*) of biotreated fabrics, for a given protease, decreased, when enzyme loading was increased (Table 8). The silk fabrics treated with papain showed 5-63% decrease in *G* values for the different enzyme activities, while the fabrics treated with Esperase® 8.0L showed a higher decrease in *G* values which ranged between 65-75%. Lower *G* values, of the enzymatically treated silk fabrics is "translated" in softer fabrics with better drape (Table 8).

3.2.6. Tensile Property

The tensile behaviour of fabric is closely related to the inter-fiber friction effect, the ease of crimp removal and the load-extension properties of the yarns themselves. Through the KES apparatus the EMT parameter was determined which reflects fabric's extensibility, a measure of a fabric's ability to be stretched under tensile load. The larger the EMT, the more extensible the fabric.

Table 8. Comparison of micromechanical properties of silk fabrics treated for 60 min with various levels of papain and Esperase® 8.0L, with reference and conventionally degummed materials

Type of treatment	Bending rigidity (*B*) (gf·cm²·cm⁻¹)			Shear stiffness (*G*) (gf·cm⁻¹·degree⁻¹)		
	warp	weft	*mean*	warp	weft	*mean*
No enzyme	0.0205±0.0005	0.0117±0.0001	*0.0161*	1.80±0.02	1.60±0.01	*1.70*
P 25 U·g⁻¹ fabric	0.0256±0.0007	0.0147±0.0003	*0.0202*	1.70±0.03	1.53±0.02	*1.62*
P 50 U·g⁻¹ fabric	0.0356±0.0009	0.0185±0.0004	*0.0271*	1.30±0.01	1.25±0.01	*1.27*
P 75 U·g⁻¹ fabric	0.0463±0.0006	0.0234±0.0002	*0.0349*	0.75±0.01	0.50±0.01	*0.63*
E 25 U·g⁻¹ fabric	0.0337±0.0004	0.0126±0.0003	*0.0232*	0.85±0.01	0.35±0.01	*0.60*
E 50 U·g⁻¹ fabric	0.0321±0.0006	0.0283±0.0002	*0.0302*	0.55±0.01	0.35±0.01	*0.45*
E 75 U·g⁻¹ fabric	0.0479±0.0007	0.0283±0.0005	*0.0381*	0.50±0.01	0.35±0.01	*0.43*
Marseille soap	0.0537±0.0006	0.0264±0.0003	*0.0401*	1.80±0.03	1.60±0.01	*1.70*

P: Papain, E: Esperase® 8.0L

Tensile property increased when enzyme concentration increased for both protease used (Table 9). Treatment with Esperase® 8.0L and particularly with high concentrations resulted in silk fabrics more elastic compared to those treated with Marseille soap (Table 9).

Table 9. Comparison of micromechanical properties of silk fabrics treated for 60 min with various levels of papain and Esperase® 8.0L, with reference and conventionally degummed materials

Type of treatment	Tensile (%)		
	warp	weft	*mean*
No enzyme	5.15 ± 0.2	1.63 ± 0.07	*3.39*
P 25 $U g^{-1}{}_{fabric}$	4.09 ± 0.3	1.24 ± 0.04	*2.67*
P 50 $U g^{-1}{}_{fabric}$	3.97 ± 0.2	0.94 ± 0.02	*2.53*
P 75 $U g^{-1}{}_{fabric}$	4.53 ± 0.3	1.33 ± 0.04	*2.93*
E 25 $U g^{-1}{}_{fabric}$	4.51 ± 0.3	1.05 ± 0.04	*2.78*
E 50 $U g^{-1}{}_{fabric}$	5.37 ± 0.1	1.49 ± 0.02	*3.43*
E 75 $U g^{-1}{}_{fabric}$	6.32 ± 0.3	1.15 ± 0.01	*3.74*
Marseille soap	4.29 ± 0.2	1.42 ± 0.07	*2.86*

P: Papain, E: Esperase® 8.0L

3.3. Synergistic Action Of Protease-Lipase On Silk Degumming: Effect Of Treatment Time

Proteases and lipases are normally used in combination for degumming and removing others impurities such as waxes, fats, mineral salts and pigments [49]. Waxes and fats as well as colorants and mineral components occur exclusively in silk gum layer (sericin) [66]. The combined effect of enzyme activity (expressed as Units of protease per g of silk fabric) and treatment time on physicochemical and low-stress mechanical properties of cotton fabric was investigated. Raw silk fabrics were treated with two different proteases namely papain and Esperase® 8.0L at an enzyme activity of 50 and 75 $U{\cdot}g^{-1}{}_{fabric}$combined with a lipase (Lipolase® Ultra 50T) at a constant activity of 50 $U{\cdot}g^{-1}{}_{fabric}$ ic for 30, 60, 90 min. Furthermore, buffer treated and conventional degummed silk fabrics were used as controls.

3.3.1. Weight Loss

The results are depicted in Figure 4 both as weight loss (Figure 4a) and degumming efficiency (Figure 4b). Without enzymes, the weight loss was negligible.

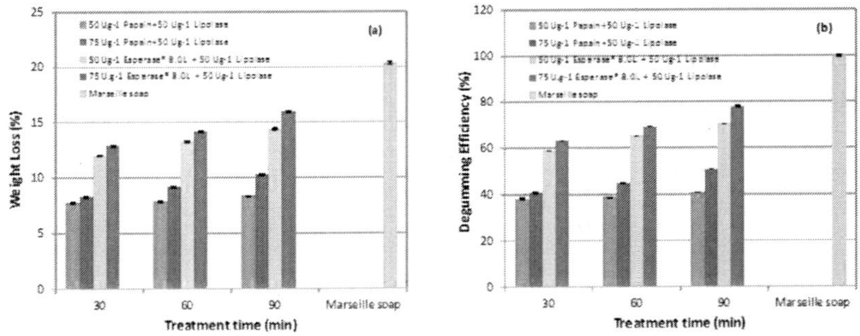

Figure 4. a) weight loss of silk fabrics treated with mixtures of proteases/lipase for 30, 60 and 90 min and (b) degumming efficiency of the enzymatic process

The degree of fabric's degradation was found to be affected by both enzyme concentration and treatment time. Increasing enzyme concentration and treatment time resulted in increased weight loss values. The combination of Esperase® 8.0L with Lipolase® Ultra 50T was found superior to that of papain. The highest weight loss (15.9%, w/w) was achieved when fabrics were treated with 75 $U \cdot g^{-1}_{fabric}$ Esperase® 8.0L + 50 $U \cdot g^{-1}_{fabric}$ Lipolase® Ultra 50T for 90min, while weight loss during conventional degumming was higher (20.4 %, w/w).

Degumming efficiency did not exceed 78% (Figure 4b). The proteases and lipase combination did not improve the results obtained when the fabrics degummed only with proteases.

3.3.2. Wettability and Crystallinity Index

Wettability seemed to depend more on the enzymes used for degumming and less on treatment time. (Table 10). The most effective combination is that of Esperase® 8.0L with Lipolase® Ultra 50T. This combination at the highest enzyme loading used, exhibited adequate water absorbency (<1 sec) after 60 min of treatment. The mixture of papain with Lipolase® Ultra 50T did not improve the wettability of the silk fabrics compared to those treated only with papain, even at higher treatment times.

Crystallinity index increased with treatment time for all enzyme combination tested (Table 10). The combination of Esperase® 8.0L with lipase was more effective compared to that of papain. However, the addition of lipase did not seem to improve Crystallinity Index in any condition tested, compared to the action of single protease.

Table 10. Wetting time and Crystallinity Index of silk fabrics treated conventionally (Marseille soap) or enzymatically.

Treatment	Wetting time (sec)	Crystallinity Index (%)
50 U g^{-1} $_{fabric}$ Papain + 50 U g^{-1} $_{fabric}$ Lipolase® Ultra 50T		
30 min	5.90 ± 0.18	62.80 ± 1.30
60 min	5.80 ± 0.12	63.20 ± 1.24
90 min	5.60 ± 0.10	63.32 ± 1.31
75 U g^{-1} $_{fabric}$ Papain + 50 U g^{-1} $_{fabric}$ Lipolase® Ultra 50T		
30 min	4.72 ± 0.20	62.90 ± 1.36
60 min	3.60 ± 0.15	63.20 ± 1.40
90 min	3.30 ± 0.30	63.32 ± 1.26
50 U g^{-1} $_{fabric}$ Esperase® 8.0L + 50 U g^{-1} $_{fabric}$ Lipolase® Ultra 50T		
30 min	2.20 ± 0.20	64.80 ± 1.60
60 min	2.00 ± 0.19	65.00 ± 1.80
90 min	1.50 ± 0.18	65.10 ± 1.10
75 U g^{-1} $_{fabric}$ Esperase® 8.0L + 50 U g^{-1} $_{fabric}$ Lipolase® Ultra 50T		
30 min	1.47 ± 0.09	65.20 ± 1.50
60 min	<1	65.40 ± 1.40
90 min	<1	66.00 ± 1.10
Marseille soap	<1	66.48 ± 1.35
No enzyme	6.85 ± 0.20	63.90 ± 1.27

3.3.3. Whiteness And Whiteness After Bleaching With H_2O_2

The synergistic action of protease-lipase improved the whiteness of the silk fabrics (Figure 5a). Increasing enzyme concentration and treatment time resulted in increased whiteness values. The highest whiteness values for each combination were observed at highest enzyme loading and after 90 min of treatment. Those values were 55.2 and 59.8 Berger degree for papain+ Lipolase® Ultra 50T and Esperase® 8.0L+ Lipolase® Ultra 50T, respectively. It should be noted that the conventional degumming resulted in a lower whiteness value (58.8 Berger degree) (Figure 5a). The addition of lipase improved the whiteness of the silk fabrics compared to results obtained by the use of protease only. For example the whiteness of the silk fabrics treated with 75 $U{\cdot}g^{-1}_{fabric}$ Esperase® 8.0L for 60 min was found 52.2 Berger degree, while the corresponding value of the fabrics treated at the same conditions adding 50 $U{\cdot}g^{-1}_{fabric}$ Lipolase® Ultra 50T was 57.3 Berger degree.

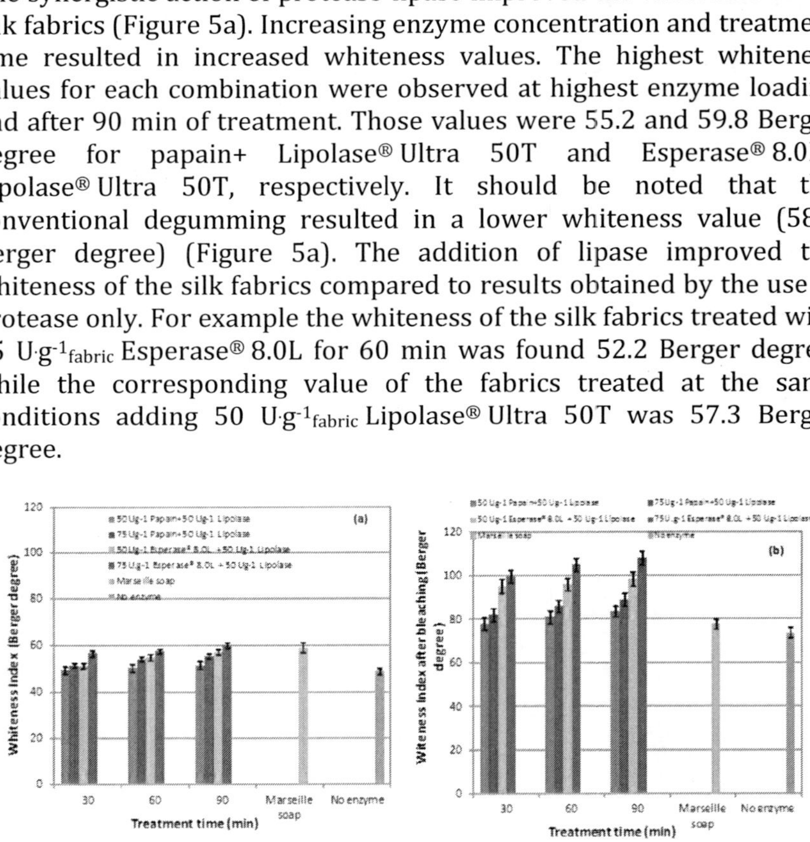

Figure 5. a) Whiteness Index and (b) Whiteness Index after bleaching with H_2O_2 of degummed silk fabrics using Marseille soap (conventional treatment) or mixture of papain and Esperase® 8.0L with Lipolase® Ultra 50T at different loadings and treatment times.

The synergistic action of protease-lipase improved the bleaching effect of the silk fabrics (Figure 5b). Increasing enzyme concentration and treatment time resulted in increased whiteness values after bleaching.

The whiteness after bleaching of the silk fabrics treated with papain and Lipolase® Ultra 50T ranged between 78.0-89.0 Berger degree, while the corresponding values for the fabrics treated only with papain were found between 68.4-76.0 Berger degree. Higher whiteness values were observed for the fabrics treated with the mixture of Esperase® 8.0L with Lipolase® Ultra 50T. The whiteness ranged between 95.23-108 Berger degree. The silk fabrics treated with Esperase® 8.0L + Lipolase®Ultra 50T exhibited 40% increase in whiteness in comparison with those treated with the conventional method with Marseille soap, while the corresponding increase in whiteness for the fabrics treated with papain and Lipolase® Ultra 50T was 15%.

3.3.4. Bending Property

Enzymatically and soap treated silk fabrics exhibited higher *B* values compared to untreated silk fabrics. Bending rigidity of the enzymatically treated silk fabrics was affected by enzyme concentration and treatment time. Increasing enzyme concentration and treatment time resulted in lower *B* values, namely in less rigid fabrics compared to conventionally treated ones (Table 11).

One interesting aspect is that addition of Lipolase® Ultra 50T caused further decrease in *B* values compared to those obtained when silk fabrics were treated only with protease. For example, silk fabrics treated with 75 $U{\cdot}g^{-1}_{fabric}$ Esperase® 8.0L exhibited a *B* value of 0.0381 ($gf{\cdot}cm^2{\cdot}cm^{-1}$) after 60 min (Table 8). Addition of 50 $U{\cdot}g^{-1}_{fabric}$ Lipolase® Ultra 50T resulted in 43% decrease in *B* value (0.0291, $gf{\cdot}cm^2{\cdot}cm^{-1}$). The same pattern was observed for all conditions tested.

3.3.5. Shear Property

Combined action of proteases with Lipolase resulted in silk fabrics with lower *G* values compared to untreated and conventional treated silk fabrics (Table 11). Increasing protease concentration and treatment time resulted in silk fabrics less rigid with better drape (lower *G* values) compared to those treated with Marseille soap.

Table 11. Comparison of micromechanical properties of silk fabrics treated at varying durations with various levels of Papain and Esperase® 8.0L combined with constant level of Lipolase® Ultra 50T, with reference and conventionally degummed materials

Type of treatment	Bending rigidity (*B*) (gf·cm²·cm⁻¹)			Shear stiffness (*G*) (gf·cm⁻¹·degree⁻¹)		
	warp	weft	*mean*	warp	weft	*mean*
No enzyme	0.0205±0.0005	0.0117±0.0001	*0.0161*	1.80±0.02	1.60±0.01	*1.70*
50 Ug⁻¹ Papain + 50 Ug⁻¹ Lipolase® Ultra 50T						
30 min	0.0300±0.0003	0.0200±0.0004	*0.0250*	1.50±0.02	1.34±0.01	*1.42*
60 min	0.0250±0.0005	0.0170±0.0002	*0.0210*	1.44±0.03	1.30±0.03	*1.37*
90 min	0.0200±0.0002	0.0160±0.0002	*0.0180*	1.28±0.02	1.22±0.02	*1.25*
75 Ug⁻¹ Papain + 50 Ug⁻¹ Lipolase® Ultra 50T						
30 min	0.0220±0.0002	0.0180±0.0001	*0.0200*	1.34±0.02	1.30±0.01	*1.32*
60 min	0.0170±0.0003	0.0140±0.0002	*0.0155*	1.29±0.03	1.25±0.01	*1.27*
90 min	0.0160±0.0002	0.0130±0.0003	*0.0145*	1.26±0.01	1.23±0.02	*1.25*
50 Ug⁻¹ Esperase® 8.0L + 50 Ug⁻¹ Lipolase® Ultra 50T						
30 min	0.0424±0.0009	0.0233±0.0002	*0.0329*	1.85±0.04	1.50±0.05	*1.68*
60 min	0.0341±0.0005	0.0200±0.0001	*0.0271*	1.80±0.04	1.45±0.05	*1.63*
90 min	0.0330±0.0007	0.0187±0.0003	*0.0259*	1.60±0.06	1.40±0.05	*1.50*
75 Ug⁻¹ Esperase® 8.0L + 50 Ug⁻¹ Lipolase® Ultra 50T						
30 min	0.0491±0.0005	0.0231±0.0002	*0.0361*	1.90±0.06	1.20±0.04	*1.55*
60 min	0.0242±0.0003	0.0195±0.0005	*0.0219*	1.58±0.03	0.90±0.03	*1.24*
90 min	0.0214±0.0007	0.0170±0.0003	*0.0192*	1.53±0.02	0.10±0.02	*0.81*
Marseille soap	0.0537±0.0006	0.0264±0.0003	*0.0401*	1.80±0.03	1.60±0.01	*1.70*

Shear rigidity was also affected by the addition of Lipolase® Ultra 50T. For example, silk fabrics treated with 75 $U{\cdot}g^{-1}_{fabric}$ Esperase® 8.0L exhibited a *G* value of 0.43 (gf·cm⁻¹·degree⁻¹) after 60 min (Table 8). Addition of 50 U/g Lipolase® Ultra 50T resulted in 65% increase in *G* value (1.24, gf·cm⁻¹·degree⁻¹). The same pattern was observed for all conditions tested.

3.3.6. Tensile Property

The elasticity, (tensile strength) of the silk fabrics treated with papain and Lipolase® Ultra 50T was higher than the elasticity of the untreated and soap treated silk fabrics and seemed to be dependent on protease dosage and treatment time (Table 12). Increasing papain dosage and treatment time resulted in an increase elasticity of the enzymatically treated of silk fabrics. At highest papain dosage and for treatment times 60 min and more, the elasticity was superior to that of Marseille soap treated silk.

Opposite trend was observed for the fabrics treated with Esperase® 8.0L + Lipolase® Ultra 50T combination. Increasing protease dosage and treatment time the silk fabrics became more anelastic (Table 12).

Table 12. Comparison of micromechanical properties of silk fabrics treated for different times with various levels of papain and Esperase® 8.0L combined with constant level of Lipolase® Ultra 50T, with reference and conventionally degummed materials

Type of treatment	Tensile (%)		
	warp	weft	*mean*
No enzyme	5.15 ± 0.2	1.63 ± 0.07	*3.39*
50 $U g^{-1}_{fabric}$ Papain + 50 $U g^{-1}_{fabric}$ Lipolase® Ultra 50T			
30 min	3.30 ± 0.10	2.70 ± 0.20	*3.00*
60 min	3.60 ± 0.10	3.00 ± 0.10	*3.30*
90 min	3.90 ± 0.20	3.10 ± 0.10	*3.50*
75 $U g^{-1}_{fabric}$ Papain + 50 $U g^{-1}_{fabric}$ Lipolase® Ultra 50T			
30 min	4.10 ± 0.20	3.30 ± 0.20	*3.70*
60 min	4.20 ± 0.20	3.40 ± 0.30	*3.80*
90 min	4.40 ± 0.40	3.50 ± 0.20	*3.95*
50 $U g^{-1}_{fabric}$ Esperase® 8.0L + 50 $U g^{-1}_{fabric}$ Lipolase® Ultra 50T			
30 min	4.85 ± 0.10	1.74 ± 0.05	*3.30*
60 min	4.75 ± 0.20	1.65 ± 0.02	*3.20*
90 min	4.60 ± 0.20	1.50 ± 0.03	*3.05*
75 $U g^{-1}_{fabric}$ Esperase® 8.0L + $U g^{-1}_{fabric}$ Lipolase® Ultra 50T			
30 min	4.96 ± 0.10	1.87 ± 0.09	*3.42*
60 min	4.62 ± 0.40	1.55 ± 0.05	*3.09*
90 min	4.51 ± 0.30	1.36 ± 0.04	*2.94*
Marseille soap	4.29 ± 0.20	1.42 ± 0.02	*2.86*

Addition of Lipolase® Ultra 50T in papain degumming mixture produces more elastic fabrics compared to those obtained by single papain treatement. For example silk fabrics treated with papain at 75 $U{\cdot}g^{-1}{}_{fabric}$ fabric for 60 min exhibited a tensile strength value of 2.93%. Addition of 50 $U{\cdot}g^{-1}{}_{fabric}$Lipolase® Ultra 50T increases this value at 3.80%. On the other hand the use of Esperase® 8.0L + Lipolase® Ultra 50T mixture causes a decrease in the fabric's elasticity.

4. CONCLUSIONS

Enzymatic degumming of silk fabric using proteolytic enzymes (papain, Esperase® 8.0L) and mixtures of thereof with a lipolytic (Lipolase® Ultra 50T) enzyme under mild conditions was investigated. The results obtained are encouraging in comparison with those of the conventional method of degumming (Marseille soap).

Silk fabrics treated with Esperase® 8.0L (an alkaline protease) at high concentration (75 U/g fabric) for 60 min exhibited degumming efficiency nearly 99%, which indicate almost complete removal of sericin from the surface of the fabric. Esperase® 8.0L treatment resulted in fabrics with adequate wettability, higher CrI and whitenees after bleaching compared to those treated conventionally (Marseille soap).

On the other hand papain was not so effective in the degumming process. This could probably be attributed to the different substrate specificity of the proteases, that is, the chemical structure of the target cleavage site. The silk fabrics treated with both proteolytic enzymes exhibited low bending and shear rigidity and higher elasticity compared to Marseille soap treated silk fabrics. This means that the silk fabrics after enzymatic degumming were softer, less rigid, with better drape and higher elasticity compared to those treated with Marseille soap.

Since the silk of *Bombyx mori* apart of the proteins fibroin and sericin, also contains fats, wax etc., the combined effect of proteolytic enzymes with a lipolytic one was investigated. Combined action of protease with lipase (Lipolase® Ultra 50T) resulted in lower degumming efficiency compared (under the same conditions) with protease treatment only, but generated silk fabrics with significant improvement in whiteness after bleaching. As far as the properties measured in the Kawabata evaluation system the combination of proteolytic with lipolytic enzymes resulted in silk fabrics with extremely low bending rigidity, reduced shear stiffness and with higher elasticity compared to Marseille soap treated. This means that the fabrics were softer, less rigid with better drape compared to conventionally treated. Addition of lipolase caused decrease in bending and increase in shear rigidity and elasticity of the silk fabrics compared to those treated only with protease.

Silk degumming is a high resource consuming process as far as water and energy are concerned. Moreover, it is ecologically questionable for the high environmental impact of effluents. The development of an effective degumming process based on enzymes as active agents would entail savings in terms of water, energy, chemicals, and effluent treatment. This could be made possible by the milder treatment conditions, the recycling of processing water, the recovery of valuable by-products such as sericin peptides, and the lower environmental impact of effluents.

REFERENCES

1. The Silk Association of Great Britain. http://www.silk. org. uk/ history.php (accessed 30 March 2012)

2. FAO: Food and Agriculture Organization of the United Nations. www.fao.org (accessed 25 March 2012)

3. Jolly MS, Sen SK. Patterns of follicular imprints in egg shell-a species-specific character in Antheraea. Bull Entomol 1969;10 32-38.

4. Jolly MS,Narasimhanna MN,Wartkar VB. Arrangement of tubercular setae in Antheraea species (Lepidoptera: Saturniidae). Entomologist 1970;103 18-23.

5. Souche YS, Patole M. Sequence analysis of mitochondrial 16S ribosomal RNA gene fragment from seven mosquito species. Journal of Biosciences 2000;25 61-66.

6. Tan YD, Wan C, Zhu Y, Lu C, Xiang Z, Deng HW. An amplified fragment length polymorphism map of the silkworm. Genetics 2001;157 1227-1284.

7. Mahendran B, Padhi BK, Ghosh SK, Kundu SC. Genetic variation in ecoraces of tropical tasar silkworm, Antheraea mylitta D. using RFLP technique. Current Science 2005;90 100-103.

8. Mahendran B, Acharya C, Dash R, Ghosh SK, Kundu SC. Repetitive DNA in tropical tasar silkworm Antheraea mylitta. Gene 2006;370 51-57.

9. Li A, Zhao Q, Tang S, Zhang Z, Pan S, Shen G. Molecular phylogeny of the domesticated silkworm, Bombyx mori, based on the sequences of mitochondrial cytochrome b genes. Journal of Genetics 2005;84 137-142.

10. http://www.csb.gov.in/silk-sericulture/silk/ (accessed 30 March 2012)

11. Kundu SC, Dash BC, Dash R, Kaplan DL. Natural protective glue protein, sericin bioengineered by silkworms: Potential for biomedical and biotechnological applications. Progress in Polymer Science 2008;33 998-1012.

12. Sinohara H, Asano Y, Fukui A. Glycopeptides from silk fibroin of Bombyx mori. Biochimica et Biophysica Acta 1971;237 273-279.

13. Prudhomme J-C, Couble P., Garel J-P. and Daillie J. Silk synthesis. In: Kerkut GA, Gilbert LJ. (eds) Comprehensive Insect Physiology, Biochemistry and Pharmacology. Oxford: Pergamon Press; 1985. Vol 10 p571-594.

14. Inoue S, Tanaka K, Arisaka F, Kimura S, Ohtomo K, Mizuno S. Silk fibroin of Bombyx mori is secreted, assembling a high molecular mass elementary unit consisting of H-chain, L-chain, and P25, with a 6:6:1 molar ratio. Journal of Biological Chemistry 2000;275(51) 40517-40528.

15. Zhou CZ, Confalonieri F, Jacquet M, Perasso R, Li ZG, Janin J. Silk fibroin: structural implications of a remarkable amino acid sequence. Proteins:Structure, Function and Genetics 2001;44 119-122.

16. Vepari C, Kaplan DL. Silk as a biomaterial. Progress in Polymer Science 2007;32 991-1007.

17. Li G, Liu H, Li TD, Wang J. Surface modification and functionalization of silk fibroin fibers/fabric toward high performance applications. Materials Science and Engineering C 2012;32 627-636.

18. Teramoto H, Kakazu A, Asakura T. Native structure and degradation pattern of silk sericin studied by 13C NMR spectroscopy. Macromolecules 2006;39 6-8.

19. Shao ZZ, Vollrath F. Surprising strength of silkworm silk. Nature 2002;418 741.

20. Jin HJ, Kaplan DL. Mechanism of silk processing in insects and spiders. Nature 2003;424 1057-1061.

21. Sinohara H. Glycopeptides isolated from sericin of the silkworm, Bombyx mori. Comparative Biochemistry and Physiology Part B: Comparative Biochemistry 1979;63(1) 87-91.

22. Vollrath F, Knight DP. Liquid crystalline spinning of spider silk. Nature 2001;410 541-548.

23. Sprague KU. Bombyx mori silk proteins: Characterization of large polypeptides. Biochemistry 1975;14 925-931.

24. Takasu Y, Yamada H, Tsubouchi K. Isolation of three main sericin components from the cocoon of the silkworm, Bombyx mori. Bioscience Biotechnology and Biochemistry 2002;66 2715-2718.

25. Colomban P., Dinh HM, Riand J, Prinsloo LC, Mauchamp B. Nanomechanics of single silkworm and spider fibres: a Raman and micro-mechanical in situ study of the conformation change with stress. Journal of Raman Spectroscopy 2008;39(12) 1749-1764.

26. Freddi G, Pessina G, Tsukada M. Swelling and dissolution of silk fibroin (Bombyx mori) in N-methyl morpholine N-oxide. International Journal of Biological Macromolecules 1999;24(2-3) 251-263.

27. Gulrajani ML. Degumming of silk. Review of Progress in Coloration 1992;22 79-89.

28. Rajasekhar A, Ravi V, Reddy MN, Rao KRSS. Thermostable Bacterial Protease - A New Way for Quality Silk Production. International Journal of Bio-Science and Bio-Technology 2011;3(2) 43-58.

29. Uddin K, Hossain S. A comparative study on silk dyeing with acid dye and reactive dye. International Journal of Engineering & Technology IJET-IJENS 2010;10(6) 21-26.

30. Wray LS, Hu X, Gallego J, Georgakoudi I, Omenetto FG, Schmidt D, Kaplan DL. Effect of processing on silk-based biomaterials: Reproducibility and biocompatibility. Journal of Biomedical Materials Research B: Applied Biomaterials 2011;99B(1) 89-101.

31. Rajkhowa R, Wang LJ, Kanwar JR, Wang XG. Molecular weight and secondary structure change in eri silk during alkali degumming and powdering. Journal of Applied Polymer Science 2011;119 1339-1347

32. Svilokos Bianchi A, Colonna GM. Developments in the degumming of silk. Melliand Textilberichte 1992;73 68-75.

33. Freddi G, Mossotti R, Innocenti R. Degumming of silk fabric with several proteases. Journal of Biotechnology 2003;106(1) 101-112.

34. Freddi G, Allara G, Candiani G. Degumming of silk with tartaric acid. Journal of the Society of Dyers and Colourists 1996;112 191-195.

35. Gulrajani ML, Chatterjee A. Degumming of silk with oxalic acid. Indian Journal of Fibre and Textile Research 1992;17 1.

36. Gulrajani ML, Sethi S, Guptha S. Some studies in degumming of silk with organic acids. Journal of the Society of Dyers and Colourists 1992;108(2) 79-86.

37. Khan MR, Tsukada M, Gotoh Y, Morikawa H, Freddi G, Shiozaki H. Physical properties and dyeability of silk fibers degummed with citric acid. Bioresource Technology 2010;101 8439-8445.

38. Vohrer U, Müller H, Oehr C. Glow-discharge treatment for the modification of textiles. Surface and Coatings Technology 1998;98(1-3) 1128-1131.

39. Borcia G, Anderson CA, Brown NMD. Surface treatment of natural and synthetic textiles using a dielectric barrier discharge. Surface and Coatings Technology 2006;201 3074-3081.

40. Yip J, Chan K, Sin KM, Lau KS. Low temperature plasma-treated nylon fabrics. Journal of Materials Processing Technology 2002;123(1) 5-12.

41. Long JJ, Wang HW, Lu TQ, Tang RC, Zhu YW. Application of Low-Pressure Plasma Pretreatment in Silk Fabric Degumming Process. Plasma Chemistry and Plasma Processing 2008:28 701-713.

42. Mahmoodi NM, Moghimi F, Arami M, Mazaheri F. Silk Degumming Using Microwave Irradiation as an Environmentally Friendly Surface Modification Method. Fibers and Polymers 2010;11(2) 234-240.

43. Mahmoodi NM, Arami M, Mazaheri F, Rahimi S. Degradation of sericin (degumming) of Persian silk by ultrasound and enzymes as a cleaner and environmentally friendly process. Journal of Cleaner Production 2010;18 146-151.

44. Yuksek M, Kocak D, Beyit A, Merdan N. Effect of Degumming Performed with Different Type Natural Soaps and Through Ultrasonic Method on the Properties of Silk Fiber. Advances in Environmental Biology 2012;6(2) 801-808.

45. Gübitz GM, Cavaco-Paulo A. Editorial: Biotechnology in the textile industry – perspectives for the new millennium. Journal of Biotechnology 2001;89 89-90.

46. Gulrajani ML, Gupta SV, Gupta A, Suri M. Degumming of silk with different protease enzymes. Indian Journal of Fibre and Textile Research 1996;21 270-275.

47. Gulrajani ML, Sen S, Soria A, Suri M. Efficacy of proteases on degumming of dupion silk. Indian Journal of Fibre and Textile Research 1998;23, 52-58.

48. Gulrajani ML, Agarwal R, Chand S. Degumming of silk with fungal protease. Indian Journal of Fibre and Textile Research 2000b;25 138-142.

49. Chopra S, Chattopadhyay R, Gulrajani ML. Low stress mechanical properties of silk fabric degummed by different methods. Journal of the Textile Institute 1996;87 542-553.

50. Nakpathom M, Somboon B, Narumol N. Papain Enzymatic Degumming of Thai Bombyx mori Silk Fibers. Journal of Microscopy Society of Thailand 2009;23(1): 142-146.

51. Arami M, Rahimi S, Mivehie L, Mazaheri F, Mahmoodi NM. Degumming of Persian Silk with Mixed Proteolytic Enzymes. Journal of Applied Polymer Science 2007;106 267–275.

52. Gulrajani ML, Agarwal R, Grover A, Suri M. Degumming of silk with lipase and protease. Indian Journal of Fibre and Textile Research 2000a;25 69-74.

53. Fabiani C, Pizzichini M, Spadoni M, Zeddita G. Treatment of waste water from silk degumming processes for protein recovery and water reuse. Desalination 1996;105 1-9.

54. Zhang YQ. Applications of natural silk protein sericin in biomaterials. Biotechnology Advances 2002;20 91-100.

55. Vaithanomsat P, Kitpreechavanich V. Sericin separation from silk degumming wastewater. Separation and Purification Technology 2008;59 129-133.

56. Aramwit P, Siritientong T, Srichana T. Potential applications of silk sericin, a natural protein from textile industry by-products. Waste Management and Research 2012;30(3) 217-224.

57. Kocabiyik S, Erdem B. Intracellular alkaline proteases produced by thermoacidophiles: detection of protease heterogeneity by gelatin zymography and polymerase chain reaction (PCR). Bioresource Technology 2002;84 29-33.

58. Bastida A, Sabuquillo P, Armisen, P, Fernández-Lafuente R, Huguet J, Guisán JM. A single step purification, immobilization, and hyperactivation of lipases via interfacial adsorption on strongly hydrophobic supports. Biotechnology and Bioengineering 1998;58(5) 486-493.

59. Kalantzi S, Mamma D, Christakopoulos P, Kekos D. Effect of pectate lyase bioscouring on physical, chemical and low-stress mechanical properties of cotton fabrics. Bioresource Technology 2008;99(17) 8185-8192.

60. AATCC Technical Manual. AATCC: Test Method 39 1980;55 286

61. Li Y, Hardin IR. Enzymatic scouring of cotton: surfactants, agitation, and selection of enzymes. Textile Chemist and Colorist 1998;30(9) 23-29.

62. Tzanov T, Calafell M, Guenitz, GM, Cavaco-Paulo A. Bio-preparation of cotton fabrics. Enzyme and Microbial Technology 2001;29 357-362.

63. Segal L, Creely JJ, Martin AE, Conrad C.M. An empirical method for estimation the degree of crystallinity of native cellulose using the X-ray diffractometer. Textile Research Journal 1959;29(10) 786-794.

64. Houde A, Kademi A, Leblanc D. Lipases and their industrial applications. Applied Biochemistry and Biotechnology 2004;118 155-170.

65. Sonthisombat A, Speakman PT. Silk: Queen of Fibres -the Concise Story 2004 1-28. (available at http://www.en.rmutt. ac.th/ prd/ Journal/Silk_with_figuresnew.pdf)

66. Zahn H, Krasowski A. Silk. In: Ullmann's Encyclopedia of Industrial Chemistry 2002

CHAPTER 11

Preparation, Characterization and Application of Ultra-Fine Modified Pigment in Textile Dyeing

Chaoxia Wang[1] and Yunjie Yin[1]

[1] Key Laboratory of Eco-Textile, Ministry of Education, School of Textiles and Clothing, Jiangnan University, Wuxi, China

1. INTRODUCTION

Pigments are kinds of insoluble colorants used for fibers, plastics and other polymeric materials, which can retain stable chemical structure throughout the coloration process in its dispersed solution [1]. The pigments were obtained from mineral materials, vegetable materials, animal waste materials since 1200 BC. The significant development of pigment in textile field from 18th century attributed to a huge expansion in the range of synthetic pigments [2,3]. Currently, pigment perhaps is one of the most commonly and extensively colorations in fiber dyeing, due to its easy application to a variety of fibers and environmental friendly aspects [4].

As a result of the different physical and chemical characteristics between pigments and dyes, dyes are able to dissolve in water and penetrate into the substrate in the soluble form, while pigments are insoluble and difficult to be dispersed in water without the aid of dispersants [1]. The major drawback of pigments is the poor dispersing stability, which is affected by the particle size and surface charge on account of the imperfection of the dispersants [5]. Pigment dispersing system includes two main types according to the medium: solvent-based pigment and water-based pigment.

Solvent-based pigment systems, in which the organic solvent is used as dispersion medium, are more suitable for certain applications [6]. However, it is difficult to obtain pigment particles that are smaller than 100 nm in organic solvents. When the particle size of pigment is large, the interactions between pigments and organic solvents (e.g., hydrogen bonding) can disrupt the interactions with stabilizing dispersants. As a result, distinct particle aggregation that influences optical properties and viscosity occurs. To obtain a stable solvent-based pigment system, appropriate pigment amount, dispersants and solvents are the key factors [7].

Water-based pigment systems are environment-friendly pathways. Pigments are dispersed into water with the aid of auxiliaries, such as dispersants, emulsifiers, anti-setting agents, etc. Water-based pigment has been widely applied in coloration for textiles, paints, architecture, wood and so on [8,9]. But the unmodified pigment dispersion that contains pigments with large particle size is unstable and also suffers from the problems as precipitation and floating color. And then the color, fastness, handle and uniformity of the fabrics dyed with the unmodified pigment dispersion are influenced.

Ultra-fine modified pigment (UMP) is referred to nanometer or micrometer pigment particle by physical and/or chemical modifications in a composite disperse system with attaching functional groups for pigment dispersion. These functional groups often have contribution to appear the vivid color for the chromophoric groups in the pigment particle by selectively reflecting and absorbing certain wavelengths of visible light [6]. The UMP dispersion usually shows relative stability and higher color strength (K/S value) which closely approaches to that of dyes [10]. When UMP is used to color, the finely divided, insoluble particle remains throughout the coloration process [1]. Without the binder or less the amount of binder added in printing paste, the good handle will be achieved. And the better fastness is attained with the stronger attraction between the UMP and the fiber [10,11]. The UMP systems are divided into nonionic pigment dispersion, anionic pigment dispersion and cationic pigment dispersion according to the charged particles. Compared to the nonionic pigment dispersion, the pigment dispersion with ionic dispersant may produce the stronger combining power to the fabric [12,13].

The cationic disperser has been widely used in many fields. It had been reported in Japanese patents that cationic disperser was applied in electrodeposits coat. Being prepared with cationic disperser, the cationic pigment dispersion is able to decrease the migration of the pigment, and favorable dyeing deepness and color fastness are also obtained when such pigment is applied to pretreated cotton fabric and terylene-cotton fabric in pad dyeing [14]. The cationic surfactant is grafted onto the acrylic acid copolymer to attain disperser which is then applied to the anthraquinone pigment to prepare ink for ink-jet printing on paper [15]. In addition, there are many applications of cationic disperser in other fields, such as papermaking, image thermomigration, ink-jet printing paper, inorganic nano-scale powder and preparing hydrophilic ethylene copolymer dispersion. Moreover, pigment dyeing by exhaust process is possible by imparting substantivity with cationic reagents to induce the necessary affinity between pigments and fibers. The chemical modification of fiber with cationic reagents has been carried out to improve the dyeing properties [1,16].

In this chapter, the preparation of UMP, such as dispersing process, grinding process, ultrasonic wave process, microencapsulated process and microfluidics process are all summarized. And the particle size, disperse stability, Zeta potential and color properties, which are used to characterize the UMP are also analyzed. Moreover, the dyeing properties of UMP on cotton, silk, wool and acrylic yarns are mainly reviewed.

2. PREPARATION OF UMP

Different from dyes, UMPs are totally insoluble in organic mediums due to its strong intermolecular aggregation, and they are required to be finely ground and dispersed in an organic medium to warrant their gloss appearance, lighting efficiency and maximizing material utilization. In order to prepare stable UMP, the particle size of UMP needs diminishing sharply to smaller than 1 μm and the process can be achieved via dispersant, microencapsulation, grinding, ultrasonic wave, microfluidizer processes and their combined process [17-19].

The stability of UMP is determined by forces among the UMP particles, when the repulsive force is higher than attractive force, the UMP particles will stably disperse into the media, otherwise, the UMP particles will aggregate together. In UMP with polymer dispersant, the repulsion forces among particles including static-electronic and steric repulsion are produced by the adsorbed polymer. The former can be measured by the Zeta potentials. The latter is closely connected with the thickness of adsorbed polymer layer [14]. The dispersant and microencapsulation processes are usually utilized together with other methods in the preparation process.

2.1. Dispersing Process

2.1.1. Low Molecular Weight Dispersant

The UMP dispersion is a surface modification technique, typical of which is a chemical bonding of low molecular weight hydrophilic moieties on UMP surfaces. Hydrophilic moieties work by electrostatic interaction between UMP particles and vehicle solvents. Usually there are no polymeric substances, surfactants or dispersion to aid in surface-modified UMP dispersions, it is necessary to add binder resins before it is applied to textiles [20].

Low molecular weight dispersants are commonly used for wetting and dispersing the UMP particles. The effect of Triton X-100 on the colloidal dispersion stability of CuPc pigment nanoparticles was investigated by Dong et al [18]. The influence of the hydrophobic chain of quaternary dispersers on the properties of pigment dispersion was discussed by Fang et al and good dispersion effects were obtained when the hydrophobic chain were 14 or 16 [21]. The aqueous suspensions of organic pigment particles using cetyltrimethylammonium bromide (CTAB) and sodium dodecylbenzene sulfonate (SDBS) as additives were prepared by Wu et al and a uniform hydrous alumina film could be formed on the organic pigment particle surface with anion surfactant SDBS [22].

But the dispersion system only using low molecular weight dispersants is still suffered from the lack of stability for long-term storage or at high temperature [8], so additional process or dispersant are necessary.

2.1.2. Polymeric Dispersant

Polymeric dispersants are a class of specially designed, structured materials and show good properties in the stabilization of organic UMP. In aqueous media, polymeric dispersants are adsorbed onto UMP surface via anchor groups to build voluminous shells or intensify the charges around UMP surface, thereby preventing flocculation and coagulation of the UMP, which can greatly improve the stability of the UMP dispersion [23-25].

The UMP particles are usually quite hydrophobic. In order to achieve a good stabilization in aqueous UMP dispersions, many formulations have been proposed. The application of polymer surfactants in combination with ultrasonic action can significantly improve the quality of dispersed systems. Some aspects concerning UMP-polymer interaction and formation of adsorption layers under mechanical action need additional elucidation. The colloid stabilization of aqueous dispersions with polymer surfactants is believed to be a consequence of adsorption of the amphiphilic macromolecules on the particle surface resulting in mono- or multi-layers of certain structure and thickness which provide certain sterical and/or electrostatic stabilization effects.

Polymer adsorption from aqueous solution on a particle surface is a result of specific interactions of various active sites on the particle surface with corresponding sites (groups) of the macromolecule. Therefore the chemical structures of the stabilizers are believed to be adjusted to the nature of each type of the particles [26]. Fu and his coworkers reported that pigment particles with the diameter of 20-120 nm were uniformly distributing in aqueous media. –COOH of PSMA which encapsulated onto the surface of pigment would build a voluminous shell and also intensify the charges of particles, which could effectively hinder the attraction among particles [27].

2.1.3. Copolymer Dispersant

Copolymer dispersants are advantageous for providing multiple anchoring sites toward UMP surface as well as structurally more designable for solvating with the selected solvents. Polymeric structures of random, A-B block, comb-like copolymers prepared by various synthetic techniques have been employed as stabilizers against particle flocculation. However, the methods of anionic and group transfer polymerization are less appropriate since the synthesis of dispersants often involves the monomers with polar functionalities.

Copolymer dispersants are suitable for stabilizing the UMP particles against flocculation during the grinding disruption and storage. The principle for achieving a fine dispersion is a thermodynamically driven interaction among dispersant molecules, UMP particles, and solvents in a collective manner of mutual non-covalent bonding such as electrostatic charge attraction, hydrogen bonding, p-p stacking, dipole-dipole interaction, and van der Waals forces [28].

Copolymer dispersants of high molecular weight have been employed as dispersants to resolve some problems through the molecular designs with multiple anchoring functionalities for interacting with the UMP surface and simultaneously with the involved solvents. The inhomogeneity in geometric shapes of any two nanoparticles may also play an important role for excluding each other from [8]. Recent developments in living/controlled polymerization including nitroxide mediated polymerization (NMP), reversible addition-fragmentation chain transfer (RAFT), and atom transfer radical polymerization (ATRP) have been reported. The copolymers with specific functionalities can be prepared from the monomers with diversified functionalities such as C_1-C_{12} alkyl(meth)acrylate, amine-functionalized (meth)acrylate, and acid-functionalized (meth)acrylate. In addition, copolymer structures can be controlled for their molecular weight distribution and are tailored for specific UMP applications [28].

2.1.4. Siloxane Dispersant

Siloxane dispersant can make UMP well dispersed in organic binders due to their hydrophilic/ hydrophobic nature. They convert the hydrophilic surface of UMPs into hydrophobic components which make UMPs compatible with hydrophobic organic resins. Siloxane dispersants can be incorporated very easily into liquids and showed an increase of storage stability, as they tend to deposit more slowly. The dispersing extent and flowability of UMP treated with different siloxane dispersants in water medium are excellent. And siloxane also shows better affinity to UMP than ammonium and nonionic polyether [30]. Siloxane with long alkane chain shows great potential to be a new type of high performance dispersant [29].

Moreover, when the pigment powder modified by siloxane dispersant (10%) is added into the water, the treated sample (b) is able to easily be wetted and enter into the water while the untreated one (a) still floats on the water surface. It is obvious that the siloxane dispersant brings good wettability to Pigment Red CI 170.

2.2. Grinding Process

It is necessary to de-aggregate and de-agglomerate the UMP particles. This is usually accomplished by mechanical action provided by high impact mill equipment, such as the sand mill and ball mill. As the UMP powder is broken down to individual particles by mechanical shear, higher surface areas are exposed to the vehicle and larger amounts of additives are required to wet out newly formed surfaces.

The grinding process can be regarded as a de-flocculation process. In the absence of stabilizing agents, some changes such as reducing K/S value, decreasing gloss and altering rheology probably occur. Grinding process offers UMP in liquids and is suitable for discrete pass as well as for circulation operation. The product passes through a high energy grinding zone inside a grinding chamber and is reduced to the targeted particle size, down to the nanometer range if required. Using a higher specific weight grinding media is likely to reduce the bead size without losing milling energy by use of equal bead filling volumes. The milling process is improved and an optimum grinding result from both a quality and cost perspective can also be achieved by selecting the best bead material, bead size and mill speed from their dependence on the milling product properties [31].

The mechanical grinding of UMPs with the aid of a dispersant is the most convenient method used to produce UMP particles [1]. But mechanical grinding process inevitably has such defects as the relatively large particle size and broad size distribution. The typical size range for particles and/or aggregates produces by traditional mechanical grinding was from 200 nm to 1000 nm. This process often combines with the utilization of ceramic grinding media for particle size reduction. K. Hayashi et al investigated the dry grinding of UMPs with silica particles generates core-shell structures with an average size comparable to the parent silica particles (20 nm) [5].

2.3. Ultrasonic Wave Process

As an environmental and efficient approach, ultrasonic energy was firstly reported to assist various textile processes by Sokolov and Tumansky in 1941 [32]. Sound is transmitted through a medium by inducing vibrational motion of the molecules through which it is travelling. Power ultrasound induces cavitations in liquids, which is the origin of the sound effect in cleaning and in chemical processes [33]. Ultrasound has been widely applied in textiles, for example, preparing nano-scale pigments, fabric pretreatment, dyeing and finishing processes.

Power ultrasound can enhance a wide variety of chemical and physical processes, mainly due to the phenomenon known as cavitation in a liquid medium, which is the growth and explosive collapse of microscopic bubbles. Sudden and explosive collapse of these bubbles can generate hot spots, i.e. local high temperature, high pressure, shock waves, and severe shear force capable of breaking chemical bonds. Dispersion of UMPs involves the application of shear forces to the agglomerates so that they break down into primary particles [34]. If a proper amount of Ultramarine is added to water and stirred briskly, the agglomerates will simply be carried within the flow and hardly change their nature. Ultra-fine modified pigment blue FFG was mixed with dispersant and dispersed with an ultrasonic homogenizer for 30 min, and the particle size was reduced to 91 nm from the original size 220 nm [35].

2.4. Microencapsulated Process

Encapsulating UMP with various polymers is a promising approach for improving the quality of the UMP dispersion. In the last decade, many techniques have been developed for encapsulating UMP [36]. It must be noted that a successful encapsulation technique should not impair the original color appearance of UMP but enhance their dispersion stabilities. Polymeric resins in encapsulation work basically as an adsorbed surface layer around UMP particles and also a fixing agent upon being colored on a substrate. The microencapsulated UMP supplies dispersion stability, and the surface-modified UMP gives fixing on a hydrophilic body, such as the fiber in textiles [20].

The whole encapsulation process probably divides into four steps below [37]. Organic UMP are dispersed into solution of polymeric dispersant leading to the polymeric dispersant absorbing onto the UMP. The absorption auxiliary is added slowly and the copolymers precipitate and encapsulate onto the UMP surface by van der Waals forces. After the precipitate is filtered and dried, the copolymer is tightly encapsulated onto the UMP surface. The copolymer onto the encapsulated UMP is hydrolyzed, and then the UMP is dispersed uniformly in aqueous media.

Dispersion is controlled by attractive force between the UMP and the hydrophobic part of the polymer, and the stability of the particles is made both by electrostatic repulsive force between the particles in water and the polymer-polymer entropic effect. The UMP are dispersed in the solid phase, and then the encapsulation and flocculation are presented. Re-dispersion is necessary to separated microcapsules and finally the complete encapsulation in the medium occurs [20]. The main differences between the surface-modified and the microencapsulated UMP are that the hydrophilic moiety on the surface of particles, i.e. the carboxylic groups in the microencapsulated UMP and the sulfonated groups on the surface of the modified UMP, and additionally the existence of a polymer shell only in the microencapsulated UMP [20,36].

2.5. Microfluidics Process

Microfluidics with excellent dispersing and smashing effects is in favor of leading particles in liquid to reducing particle size to a submicron level to create pure and stable nano-emulsions and suspensions [38]. With this process, the effectiveness of high-performance materials increases, because particles are more uniformly and stably dispersed. Through microfluidizer high shear fluid processors, the fabrics with UMP easily achieve high K/S value and gloss, and the amount of volatile organic compounds by increasing water content is declined. Microfluidizer high shear fluid processors are frequently used to reduce particle size to less than 150 nm.

For example, Pigment Red CI 22 was stirred with an aqueous solution of an anionic polymeric dispersant for 30 min at 10,000 rpm, and the average particle diameter of the obtained UMP dispersion was 1, 500 nm. Whereas, the average particle diameter of 128 nm was available as the dispersion was converted to modified dispersion in a microfluidizer at pressure 22,000 Pa for 2.5 h [13]. Also, a UMP suspension system was prepared by adding the Gemini dispersant and water. Then it was mixed and dissolved and the deformer was added in the mixing process. The Pigment Red CI 22 was added into the above dispersant solution stirring at 600 rpm for 10 min and 9000 rpm for 30 min. The UMP system was treated using a microfluidizer at 172, 368 kPa for 50 min to prepare UMP suspension system. And particle size of the UMPs was the 211.9 nm, the Zeta potential was 32.4 mV and the viscosity was 1.33 Pa·s [16].

2.6. Combined Process

It is exceedingly difficult to handle a dry powder comprising of UMP particles as the UMP is in the form of larger soft agglomerates. These agglomerates must be broken down into the medium in a process called dispersion. The UMP dispersion can be stabilized by dispersing agents in order to prevent the formation of uncontrolled flocculates. Combining two or more dispersing methods, for example, grinding/ultrasonic wave, microencapsulation/grinding and dispersant/microfluidizer, the homogeneity of making UMP dispersion along with the use of organic dispersants for viscosity control and the prevention of particle from further agglomeration can be realized [8].

Dispersions are usually prepared by ball grinding a mixture of the pigment, dispersant and solvent. This is a simple combined process in which grinding process is used together with dispersant. The ultrasonic process can also disperse the UMP and this method can endow a high-throughput approach. The methacrylic copolymer was used to disperse carbon black in water using both ball mill and ultrasonic approaches. The particle size distribution was determined by laser diffraction. After 16 min both ultrasonification and ball grinding achieve the same particle size distribution (Figure 1). The ultrasonic approach could be used to obtain much smaller sample volumes than the ball mill approach [39].

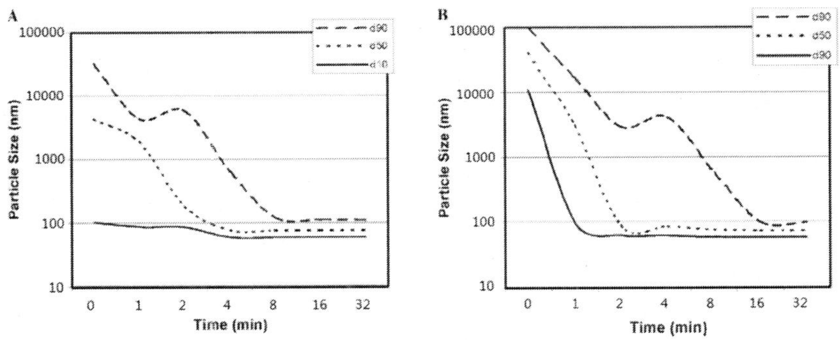

Figure 1. (A) Particle size distribution data from ball grinding; (B) Particle size distribution data from ultrasonication

3. CHARACTERIZATION OF UMP

The properties of UMP dispersion can be described by the stability, the adsorbed layer thickness and the Zeta potentials, which are greatly affected by ionic strength, dispersants and dispersing methods. The adsorbed layer thickness and the Zeta potentials can reflect the forces among polymeric dispersants, media, and UMP particles. The characteristics of UMP dispersion affect the applications of the dyed fabric on K/S value, gloss, brightness, and transparency. Some key dispersion characteristics, such as particle size, disperse stability, color properties and Zeta potential properties are investigated as follows.

3.1. Particle Size

UMPs are composed of fine particles which are normally in the submicrometer size range. The UMPs particle size has influence on the color, hide and settling characteristics. Large particles usually settle faster than smaller ones. UMP size and its distribution also influence the light scattering, colloidal stability, appearance and the color properties of the dyed fabric. Therefore, an assessment on the degree of dispersion is necessary to be considered in terms of these critical measurements. In general, color properties, such as strength, transparency, gloss and light fastness of all UMP systems, are affected to a greater or lesser extent by the size and distribution of the UMP particles in the dispersion [1].

UMP particle size and its distribution are greatly influenced by the dosage of dispersants. The particle size decreases at first and then increases with the enhancement of the dispersant, and the particle size reaches its minimum when the ratio is about 10% for phase separation method. The smaller the particle size is, the larger the UMP surface is, thus the more amount of dispersant is needed. However, excessive dispersants, higher than 10%, will dissolve in media instead of attaching onto UMP surface, resulting in increasing viscosity of dispersing media and poor wetting performance in that excess dispersants (Figure 2) [30].

The -COOH group of copolymers encapsulated onto the surface of UMPs would form a polymeric shell around the particles and also increase the surface charges of UMP particles. The interaction between UMP particles is weakened due to the existence of these charges and polymeric layer on the UMP particle surface, so that the UMP particles can be finely dispersed in aqueous media [37]. With the dosage of siloxane dispersant descending, the percentage of big particle ascends at first. However, while the dispersant is greatly excessive (10 wt%), the particle size increases.

Initial experiments, using dilute aqueous dispersions of carbon black of known average-particle size distribution, were performed using both a standard UV/Vis spectrometer and a digital camera to estimate light-transmission/attenuance. The results, summarized in Figure 3(A) and (B), showed a strong relationship between average particle size and light-transmission/attenuance over the size range examined by both these approaches, as demonstrated by the high R^2 values of the polynomial trend-line. Because the limited nature of the particle size distributions obtained during these initial experiments, we were only able to demonstrate a correlation over a small size. The digital image method was developed further on account of its potential to allow parallel analysis of multiple samples from the same image [39].

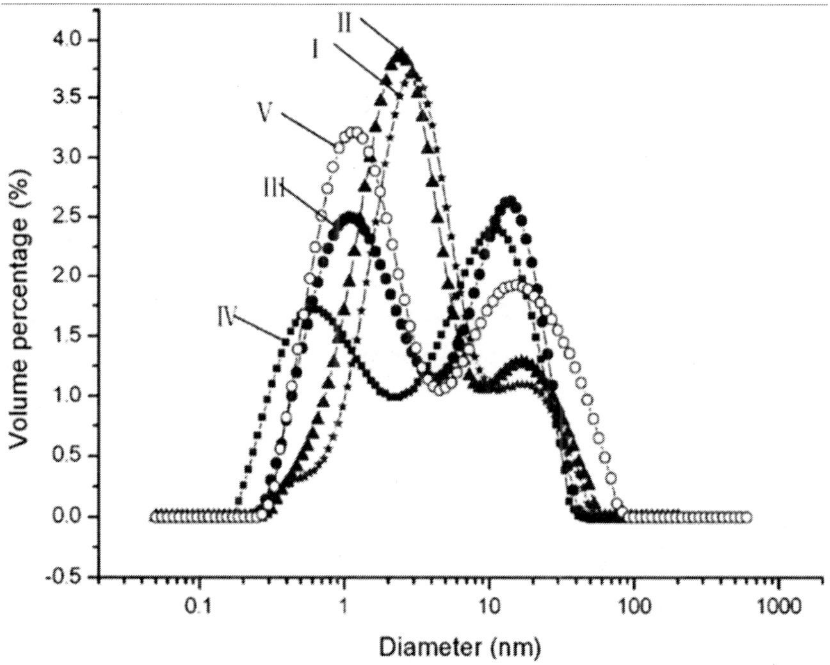

Figure 2. Particle size distribution of treated and untreated pigment red 170. (I) 1wt%pigment, 0 wt% dispersant; (II) 1wt% pigment, 0.01wt% dispersant; (III) 1wt% pigment, 0.02wt% dispersant; (IV) 1wt% pigment, 0.1wt% dispersant; (V) 1wt% pigment,10wt% dispersant.

There was a very interesting phenomenon of the particle size distribution of Pigment Red CI 170 with the increase of the cationic siloxane dispersant. In order to clearly describe the distribution process of UMP in water, a experiment was proposed according to the principle of physical chemistry. UMP could not be well dispersed in water with the absence of dispersant. Then, a small amount of dispersant was added into the system. Since the siloxane group of dispersant showed great affinity to the UMP, it tended to anchor on the surface of UMP particle.

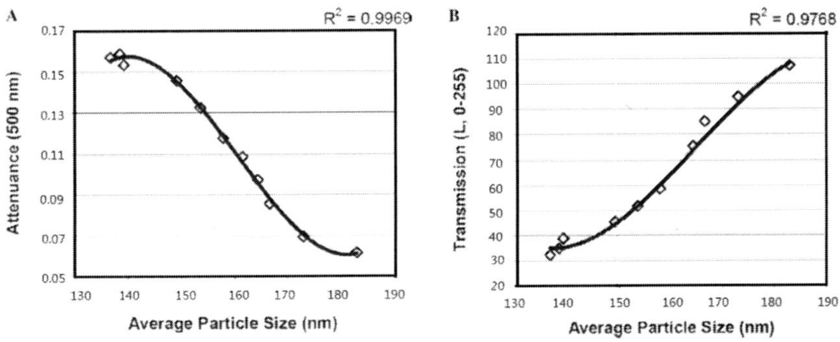

Figure 3. (A) Pigment average particle size versus attenuance (D) using UV/Vis analysis. (B) Pigment average particle size versus light-transmission using image analysis

3.2. Disperse Stability

Stability of aqueous UMP dispersion is referred to that colloidal properties such as Zeta potentials, viscosity, and storage stability will not change in certain period. It's an important performance of UMP dispersion. Disperse stability can be supported by two generally mechanisms: steric stabilization and charge stabilization. Steric stabilization is due to steric hindrance resulting from the adsorbed dispersing agent, the chains of which become solvated in the liquid medium, thus creating an effective steric barrier that prevents the other particles from approaching too close. Charge stabilization is due to electrical repulsion forces, which are the result of a charged electrical double layer surrounding the particles. The charged electrical double layer developed around the particles extends well into the liquid medium, and since all the particles are surrounded by the same charge (positive or negative), they repel each other when they come into close proximity. The disperse stability usually includes deposited stability, heat stability, pH stability, electrolyte stability and centrifugal stability [35].

Repulsive and attractive forces among UMP particles determine the stability of the UMP dispersion. When the van der Waals forces between UMP particles are higher than steric and static repulsive forces, the UMP particles will combine each other and generate large particles. Instead, the dispersion will stably exist in aqueous media. In aqueous UMP dispersion, polymeric dispersants encapsulated onto the surface of UMPs are ionized in acid solution, and produce some negative charges onto the surface of UMPs to create the static forces between UMP particles. Further, the PSMA encapsulated onto the surface of UMPs can also generate steric repulsion [27]. The whole steric and static forces are larger than van der Waals force between UMP particles, the dispersion will exist in long time, even under centrifugal force or treated at high temperature. It was reported the effect of Triton X-100 on the colloidal dispersion stability of CuPc-U (unsulfonated and hydrophobic copper phthalocyanine) particles and was concluded that the stabilization mechanism for the CuPc-U is inferred to be primarily steric and adding $NaNO_3$ had no obvious effect on the dispersion stability [18].

Stabilization of UMP with polymeric dispersants has been proven to be a good way for the preparation of UMP dispersion with high stability, small particles size, low viscosities, and low moisture sensitivity. In aqueous media, the polymeric dispersant can build polymeric shell around the UMP particles and increase the surface charge on the UMP particles, so that the UMP particles are able to finely dispersed in the aqueous media.

3.3. Zeta Potential

The Zeta potentials of UMP dispersion are analyzed to estimate their disperse stability. The stability of UMP dispersion determined by forces among the UMP particles. When the repulsive force is higher than attractive force, the UMP particles will stably disperse into the media. Otherwise, the UMP particles will arregate together. In the UMP dispersion with polymer dispersant, the repulsion forces among particles produced by the adsorbed polymer include static-electronic and steric repulsion [40]. The former can be measured by the Zeta potentials, and the latter is closely connected with the thickness of adsorbed polymer layer. The studies on the change of Zeta potentials and thickness of adsorbed polymer layer after the addition of solvent are important in understanding the effects of solvent on the stability of UMP dispersion [35,41].

The dispersion mechanism may be interpreted with a pair of electricity layers principle. Although many cationic dispersants gathered on the UMP surface neighbourhood, the quiet balance of positive negative charge didn't change. The cation on the UMP surface adsorbed a great many anions. With cation gathering gradually on the UMP surface, a pair of electricity layers formed. The particles stably existed in dispersion system for the electric charge rejected function mutually. An optimum dispersant dosage would exist, which promoted the formation of double electricity layers [16,42].

3.4. Color Properties

UMP particles which are dispersed finely in dispersions enhance the color depth of UMP dispersions (e.g. transmittance, chroma, and lightness) and improve color display performance. The transmittances were enhanced with decreasing the mean size of dispersion (Figure 4). A color analysis program estimated the lightness (L*) and chroma (C*) from a spectrum of dispersion samples at a wavelength of 380 to 780 nm based on the L*a*b* Color System and D65 light source. The finer sizes of UMP particles in dispersion resulted in the higher lightness (L*) and chroma (C*). Serious aggregation of UMP particles without a dispersant in the dispersion would be performed. Holding the supercritical fluid-assisted dispersion process at the supercritical region conduced to better dispersion [43].

Color properties are presented the color of the UMP dispersion. The absorbance of UMP hydrous dispersion is slightly larger than that of containing organic solvents. The change is mainly attributed to the disparate polarities of the solutions. In these two solutions, while the polarity of solvent UMP dispersion, which contains more ethanol, is weaker. The color of UMP is aroused by the $\pi \rightarrow \pi^*$ energy transition when the chromophore groups are irradiated. The weak ionization of the UMP increases in a solution with high polarity (such as H_2O), and the electric charge in the conjugated system can transport more easily, which leads to an increases in the absorbance. However, the maximum absorption wavelengths for both systems with or without solvent nearly remain the same. This indicates that the chromophore groups of the UMP are not damaged in the UMP dispersion containing organic solvent and the conjugated systems are essentially not altered, therefore, the color hue of the solvent UMP dispersion remains identical to that of the UMP hydrous dispersion [35].

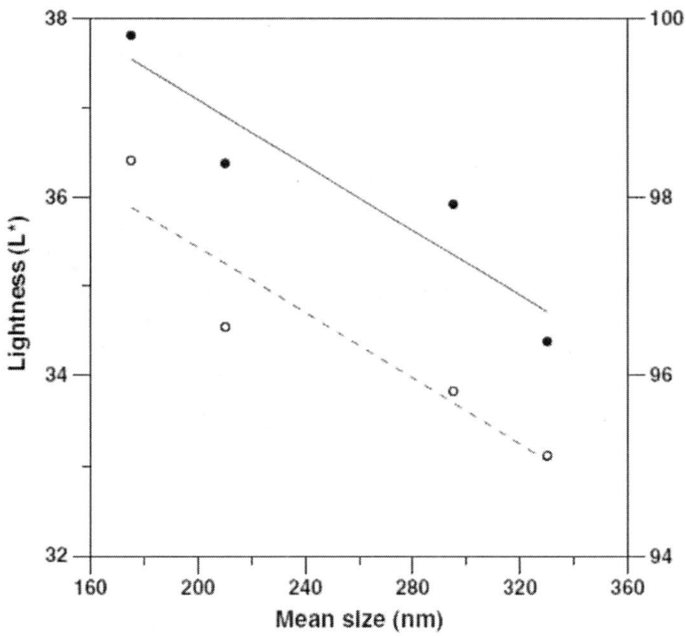

Figure 4. The color analyses from the four representative samples of dispersions. (o), Lightness; (•), chroma

4. DYEING PROPERTIES

4.1. Dyeing Cotton Fabric With Ump

In order to improve the dyeability with UMP, the cationic pretreatment of fabric is used. It can introduce positively charged sites on cotton fabric. Without pigment modification, there are two inhibiting factors. Firstly, pigments are insoluble and have no affinity for cotton and, secondly, the surfaces of pigment particles are usually negatively charged. UMP dyeing has several noticeable advantages compared to dyes. This method is a short process, simple operation, saving energy consumption and low costs, and matching intuitive easy color imitation, steady color hue, hiding power. Because UMP particles have no affinity for fibers, the UMP dyeing fibers does not exist the selective problem and is appropriate for all spices.

There are many reports on cationic modification of cotton. Karrer researched the cationic pretreatment to the cellulose fiber [44]. Guthrie studied the cationic pretreatment of cellulose fiber and dyed it with acidic dye [45]. The chemical modification of cotton to promote dyeability, light fastness and washing fastness has been researched by Cai [12]. Wang investigated chemical modification of cotton to promote fiber dyeability [46]. Burkinshaw analysized cationic pretreatment of cellulose fiber to improve the dye reactivity [47]. Hauser examined the dyeing behavior of cotton that had been rendered cationic by reaction with 2,3-epoxy propyltrimethylammonium chloride and the result showed that excellent dye yields and color fastness properties were obtained without the use of electrolytes, multiple rinsings or fixation agents [48].

During exhaustion dyeing process, the dyeing uptake of cotton fabric without modification was very low. Via cationic reagent modification, the uptake of the cationic fabric was increased significantly. Cationic reagent with higher proportions increased the number of positive charges on the cotton fiber, in turn increased pigment uptake.

For example, the cationic polymerization modifier was used to modify the charge of cotton fiber, and this could enhance the uptake of UMP and K/S value with different cationic reagent concentration (Figure 5). The application of cationic polymerization modifier on fibers for cationic modification could reduce the production cost, reduce dyeing temperature, shorten the dyeing process and improve production efficiency [49].

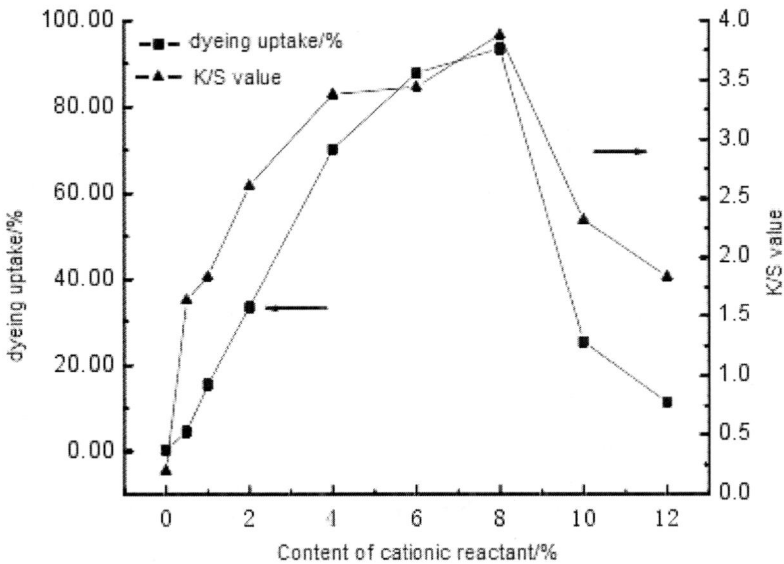

Figure 5. Effect of cationic reactant concentration on pigment exhaustion and the K/S value

4.2. Dyeing Silk Fabric With Ump

Silk is a natural protein fiber, and the shimmering appearance of silk is due to the triangular prism-like structure of the silk fiber, which allows silk cloth to refract incoming light at different. But the silk fiber also has some defects, such as the problem of bleaching, the loss of pupa protein and the yellowing [50,51].

In order to improve the performance of the silk fiber, silk was modified with cationic dispersant [52]. This improved substantivity of UMP due to the introduction of cationic groups into the silk fabric, and the balance might be due to the adsorption of cationic reagent on surface of silk fabric reaching saturation values. In this situation, more cationic reagent molecules combined with UMP. The bounds between UMP and silk fabric led to a higher K/S values. The different amount of the cationic reagent in pretreatment could impart different amount of positive charge to silk surface, which could affect the K/S values. The

UMP was not substantive to silk, the K/S values were low on silk without cationization pretreatment. The K/S values greatly increased and then kept in a balance with the increase of the cationization concentration [53].

Furthermore, penetration property and content of ammonium of silk would also restrict to produce the stronger ionic attraction between the cationic fiber and the anionic UMP. As a result, the adsorption of UMP on silk fabric did not change any more accordingly [54]. Wei reported that after modifying with glycidyltrimethylammonium chloride (EPTAC), the reactions of some amino acids were more rapid than the reaction of glucose. The dyeing ability of modified silk fiber had been improved remarkably. A theory which was called "fixed points" was given to this phenomenon [51]. Wang found that pH in pretreatment affected significantly on the dyeing properties of silk fabric with UMP, because in alkali condition, the chlorohydroxypropyl group of 3-chloro-2-hydroxypropyltrimrthylammonium chloride (CHTAC) converted to an epoxy group and then formed 2, 3-epoxypropyltrimethylammonium chloride which reacted with the nucleophilic amine group in the silk. And The highest K/S values could achieve at pH 8 [53].

With the increase of cationic reagent concentration, the rubbing fastness values appeared increasing trend (Table 1). The washing stain fastness of treated fabric was similarly comparable to the results without fastness improving reagent treatment because the UMP had no affinity with silk without treatment. Compared with untreated silk, the washing change fastnesses of cationic silk increased greatly. This was probably due to the fastness improving reagent acting as the binder between UMP and silk fabric [53].

Table 1. Color fastness at different cationic reagent concentration

Fastness improving reagent conc. (g/L)	Rubbing fastness (Grade)		Washing fastness (Grade)		
	Dry	Wet	Color change	Stain cotton	Stain silk
0	2	1-2	2	4	4
10	2-3	2	3-4	4-5	4
20	3	3	4	4-5	4
40	3-4	3	4	4-5	4-5
80	4	3-4	4	4-5	4-5

Cationic pretreatment depended on the extents of reaction between cationic reagent and silk. This might affect the physical properties of the treated fabrics. The tensile strength and elongation at break of fabrics changed with the increase of pH. It was clear that the reaction between cationic reagent and silk took place significantly when pH of the pretreatment solution increased. The tensile strength decreased with the peptide chain hydrolyzation when the pH increased. Also the elongation at break kept slightly decreasing. This revealed that the cationic pretreatment has little effect on the physical properties. The bending rigidity and hysteresis of cationization silk had no change compared with untreated silk. It revealed that the cationization treatment had no impact on the soft properties of fabrics. The handle of fabric decreased a little after dyeing, it might be due to the action of fastness improving reagent.

4.3. Dyeing Wool Fabric With Ump

The wool fabric was modified by cationic reagent and dyed with UMP which was prepared with anionic polymer disperser via exhaust method [55]. The influence of pretreatment conditions such as concentration of cationic reagent, pH value of the bath, temperature and duration of treatment on the dyeing property of the fabric was of importance to the drying properties. With the increase of the cationic content, the K/S value enhanced sharply and then kept a constant value when the content on cationic reagent was higher than 10% (Figure 6). The dry and wet rubbing fastness of the wool fabric dyed with UMP via exhaustion was gray scale ratings of 3-4 and 4, respectively. The tensile strength and elongation at break decreased slightly with the increase of bath pH value. The test results of bending rigidity and hysteresis of bending revealed that the wool fabric had soft handle and good elasticity.

4.4. Dyeing Cotton Fabric With Carbon Black Ump

The excellent color and fastness properties make carbon black (CB) particles possible application in textile dyeing. Compared with common sulfur dye, the dyeing process of carbon black UMP can save water and energy due to the shorter dyeing time. However, the dispersion of carbon black UMP in aqueous solutions and adsorption of the particles to cotton fibers are critical in exhaustion dyeing process, because carbon black UMP is hydrophobic and easily aggregate in aqueous solutions.

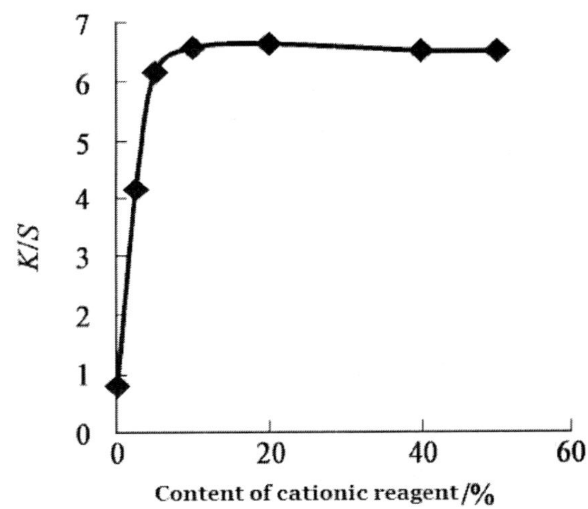

Figure 6. The contents of cationic reagent on the K/S value of dyed fabric

On the basis of cotton fabrics modified with cationic reagent, the dyeing process and properties of CB dispersions were reviewed (Figure 7). The mechanism of cationic cotton dyed by an exhaustion process using aqueous carbon black UMP dispersions was investigated. Cotton modified with a cationic reagent enhanced the dyeing properties of aqueous carbon black UMP dispersions. There were higher affinity between carbon black UMP and cationic cotton. CB particles could quickly adsorb on the surface of cationic cotton fibers within 5 min and diffuse into the inner pores and cracks on the surface layer of cationic cotton step by step [56]. The cross sections of cationic-modified cotton fibers dyed with different sizes of CB particles were colorless region, and CB particles were only in the cotton fiber grooves, just as "ring-dyeing". After cationic modification, the cotton fiber could have a dark color for the adsorbability of the anionic CB [13].

The color assessment of CB dyed cellulose is correlative to the amount carbon black UMPs on the cotton via electrostatic force. It made possible to compare the charge density difference between nanoparticle ionic materials (NIMs) deposition process and chemical modified process.

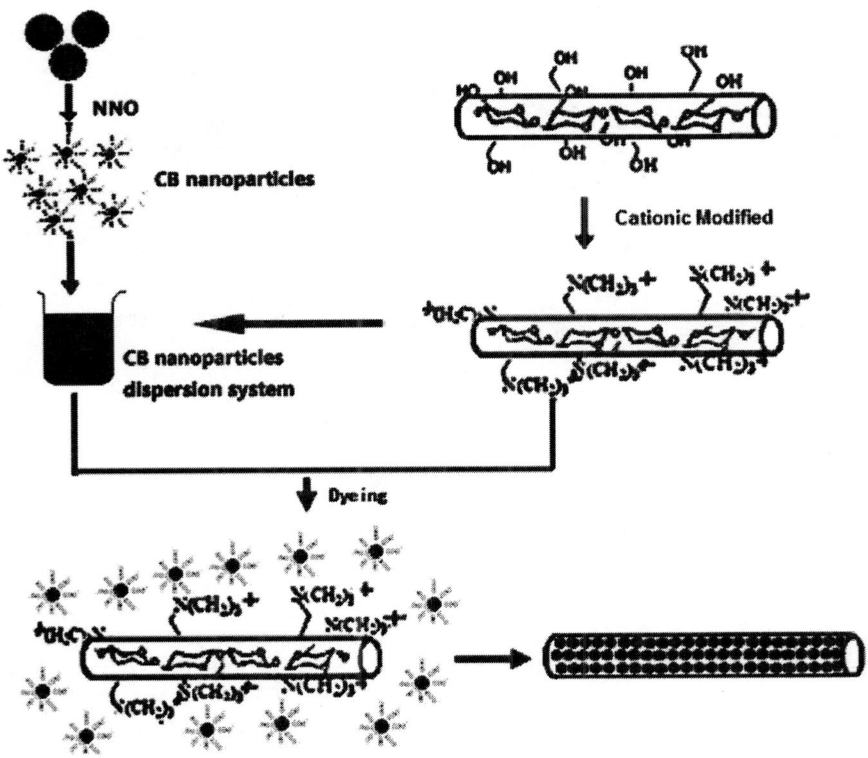

Figure 7. The dyeing process of carbon black UMP on modified cotton fabric

The results of the K/S presented that the sulfonated cellulose had the higher K/S value of 29.7 than the carboxyl of 28.6 and aminocellulose of 27.9 (Figure 8(A)). The results proved the conclusion about the color depth of the decorated samples. The images (Figure 8(B)) of the samples before and after the dyeing showed that the color depths of the dyed carboxyl and ammonium fabrics were not as deep as the sulfonated cellulose fabric, which further confirmed the color depth conclusion. Especially for the image of the ammonium sample, the fabric surface was uncoated with carbon black UMP, and there were some starred white flake on the fabric.

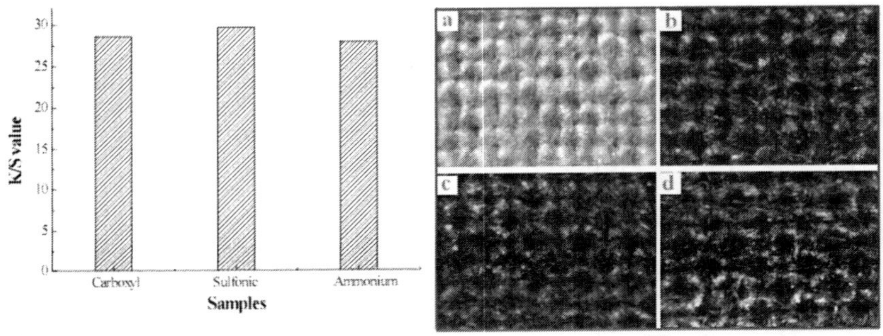

(A) (B)

Figure 8. Color assessment of CB dyed cellulose matrixes: (A)K/S values; (B) Video microscope microscope:(a) Cotton fabrics without modification, (b) Carboxyl modified cotton fabric deposited with positive NIMs and carbon black UMPs, (c) Sulfonated modified cotton fabric deposited with positive NIMs and carbon black UMPs, (d) Ammonium modified cotton fabric deposited with carbon black UMPs

4.5. Dyeing Cotton Fabric with Cationic Ump

Surface treatments are effective for UMP because its surface contain polar or polarized functional groups, which can serve as adsorption sites for the hydrophilic or lipophilic groups of the surfactants [1]. The cationic UMP dispersion has great prospects in cellulose dyeing, ink-jet printing, and so on. Compared with the anionic and non-ionic UMP, the cationic UMP has stronger combining force with cellulose substrate, better K/S value, and vividness, thus it can increase the uptake of the

UMP, reduce the environment pollution and improve the quality of final products.

For the specific structure, there are a lot of the characteristics on Gemini cationic dispersant. The Zeta potential of the Gemini cationic UMP is higher than that of the ordinary cationic UMP. Therefore, it is easier to color. The dry and wet rubbing fastness is good with both Gemini cationic UMPs after the film fixation (Table 2). With the dispersant dodecyltrimethylammonium, the dyeing of Gemini cationic UMPs exhibits deeper K/S value and the fabric is easier to be colored. The adsorption capacity of the Gemini cationic UMPs to cotton fabric is stronger, and can cover more fiber surface. Cotton fabric shows negative in a water solution, which has a weak charge bonding force with cationic UMPs. The more cationic UMPs containing electropositive, the easier the cotton fabric adsorbing the UMPs [16].

Table 2. K/S value and fastness of Gemini cationic UMP dispersion on cotton

Sample	K/S value	Dye rubbing fastness	Wet rubbing fastness
Traditional pigment	42.64	3-4	2
UMP	51.99	4	2

4.6. Dyeing Acrylic Yarns with Cationic Ump

Acrylic fibers can be dyed by cationic UMPs, and the electrostatic attraction between UMPs and acrylic fibers is beneficial to the dyeing properties, especially its light fastness property. Compared with anionic and nonionic UMP systems, the cationic UMPs could bond with the acrylic fibers by electrostatic attraction. Therefore, the UMP presented a better dyeing result and a higher utilization ratio [57].

The dye bath pH presented great influences on cationic UMP dispersion stability during dyeing acrylic fiber with cationic UMP. The increase of UMP dosage resulted in lower dye-uptake and more brilliant colors, and dyeing acrylic yarns with UMP (40%, o.w.f) could reach the maximum K/S value. The K/S value of acrylic yarns reduced after adding adhesives, compared to the cationic dyes, the light fastness was improved significantly [58].

From Figure 9, when the temperature was lower than 65 °C, the dyeing rate was quite low, and the adsorption of pigments was unstable, while the temperature was higher than 65 °C, the dyeing rate increased significantly with the dyeing time prolonging. A higher dyeing temperature would shorten the dyeing time. As the temperature was higher than 95 °C, the dyeing time was only 15 min. The reason was that in a low temperature condition, the acrylic molecular chains were not moving and the adsorption only took place in the surface of fiber. The pigments adsorbed to the yarns were easily re-dissolved into the dye bath.

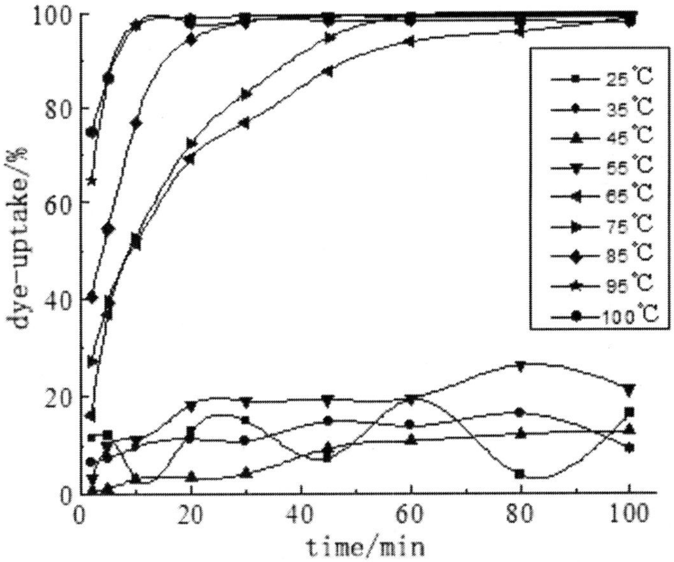

Figure 9. Effects of dyeing temperature on dye-uptake of acrylic yarns

In order to evaluate the light fastness of acrylic yarns dyed by cationic dye, the light fastness of cationic dye is attempted and assessed (Figure 10).

Cationic UMP presented better light fastness than that of the cationic dye. Acrylic yarns dyed by cationic UMP presented higher stability to light than that dyed by cationic dye. The main reason was that the chromophoric group of pigment was more difficult to destroy for its stable chemical structure.

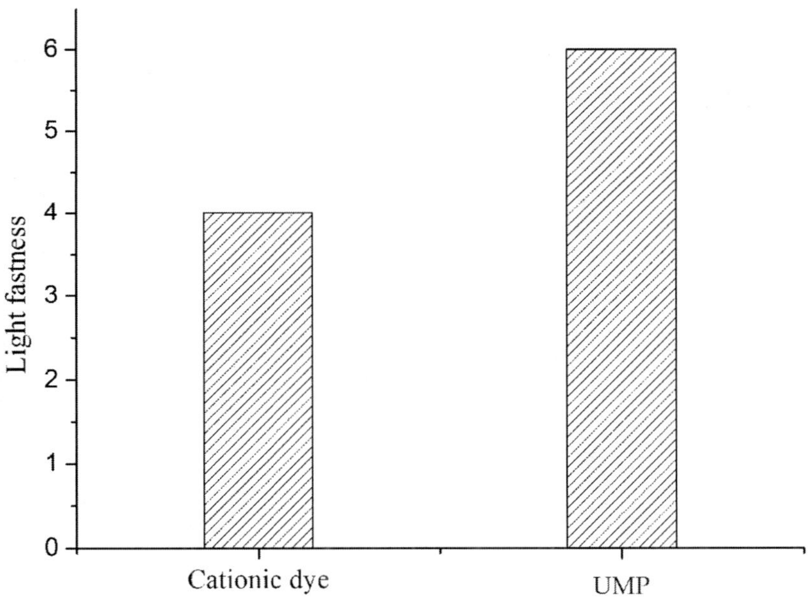

Figure 11. Light fastness of acrylic yarns dyed by cationic dye and ultra-fine pigment

5. CONCLUSION

A stable UMP dispersion is able to be prepared via grinding, ultrasonic wave, microencapsulation and microfluidics processes with different dispersants like low molecular weight dispersant, polymeric dispersant, copolymer dispersant and siloxane dispersant. It's more efficient to prepare the uniform and stable UMP dispersion via comprehensive utilization of these dispersing processes.

The key dispersion characteristics of UMP dispersion be evaluated by particle size, disperse stability, Zeta potential, color properties and dyeing properties, which are greatly affected by ionic strength, dispersants and dispersing methods. The disperse stability and color performance are promoted with the smaller particle size and higher charged UMP.

The dyeing properties of UMP on many fabrics such as cotton, silk and wool pretreated with cationic reagent is enhanced. The K/S value and rubbing fastness are able to be distinctly improved. Cationic cotton can also enhance the dyeing properties of aqueous carbon black UMP dispersions due to the higher affinity between carbon black UMP and cationic cotton. The cotton fabric dyed with cationic UMP can also promote the K/S value and color fastness. For acrylic yarns, the increase of UMP dosage results in lower dye-uptakes and more brilliant colors, and dyeing with UMP (40%, o.w.f) can reach the maximum K/S value on acrylic yarns.

As an advanced pigment dyeing method, the UMP takes many advantages of pigment. The specific characteristics of the UMP, including stability and high K/S value, reveal its significant potential on fabric dyeing, and especially for its environment-friendly characteristics, the development of UMP will present a more sharply speed than that of the dyes. Base on the textile application, the further research interests about UMP are focus on the novel dispersing process to obtain stable UMP system, modification with functional organic or inorganic materials, developing the multifunction and multipurpose of UMP.

6. ACKNOWLEDGEMENTS

The authors are grateful for the financial support of the National Natural Science Foundation of China (21174055) and the 333 Talent Project Foundation of Jiangsu Province (BRA2011184).

REFERENCES

1. T. Vernardakis, Dispersion. Pigment, 2006In: Tracton A A, editor. Coatings Technology Handbook Third Edition. Taylor & Francis Group, LLC2006. 7671

2. J. Gage, 1999Color and Culture: Practice and Meaning from Antiquity to Abstraction.University of California Press, 1218

3. Velmurugan, P., Kim, M., Park, J., Karthikeyan, K., Lakshmanaperumalsamy, P., Lee, K., Park, Y., & Oh, B.2010Dyeing of Cotton Yarn with Five Water Soluble Fungal Pigments Obtained from Five Fungi[J].Fiber Polym, 114598605

4. El -Shishtawy, R, & Nassar, S. Cationic pretreatment of cotton fabric for anionic dye and pigment printing with better fastness properties[J]. Color Technol, 2002

5. K. Hayashi, H. Morii, K. Iwasaki, S. Horie, N. Horiishi, K. Ichimura, 2007Uniformed nano-downsizing of organic pigments through core-shell structuring[J]. Journal of Materials Chemistry, 176527530

6. Kelley, A., Alessi, P., Fornalik, J., Minter, J., Bessey, P., Garno, J., & Royster, T.2010Investigation and Application of Nanoparticle Dispersions of Pigment Yellow 185 using Organic Solvents[J]. ACS Appl Mater Interfaces, 216168

7. Y. Kobayashi, H. Chiba, K. Mizusawa, N. Suzuki, J. M. Cerda-Reverter, A. Takahashi, 2011Pigment-dispersing activities and cortisol-releasing activities of melanocortins and their receptors in xanthophores and head kidneys of the goldfish Carassius auratus[J]. Gen Comp Endocr, 1733438446

8. Lan, Y., & Lin, J.2011Clay-assisted dispersion of organic pigments in water[J]. Dyes Pigment, 9012127

9. Ricaurte-Avella, G., Osorio, Velez. J., Nino, Cardona. J., Correa, J., & Alberto, Rodriguez. H.2010Evaluation of the pigment dispersion in acrylic resins for the manufacture of artificial teeth[J]. Revista Facultad De Ingenieria-Universidad De Antioquia, 53 5463

10. Ahmed, N., & El -Shishtawy, R.2010The use of new technologies in coloration of textile fibers[J]. Journal of Materials Science, 45511431153

11. Wang, C., Zhang, X., Lv, F., & Peng, L.2012Using carbon black nanoparticles to dye the cationic-modified cotton fabrics[J]. J Appl Polym Sci, 124651945199

12. Y. Cai, M. Pailthorpe, S. David, 1999A new method for improving the dyeability of cotton with reactive dyes[J]. Textile Research Journal, 696440446

13. Wang, C., & Zhang, Y.2007Effect of cationic pretreatment on modified pigment printing of cotton[J]. Materials Research Innovations, 1112730

14. Ishwarlal, R.2000Dyeing anionic cellulosic fabric materials with pigment colors having a net cationic charge using a padding process. US 27

15. T. Koyanagi, 2002Ink sets for cellulose fiber-based recording media, printing method therewith and pinted products therefrom. 23

16. C. Wang, Y. Yin, K. Fang, 2008Investigation of gemini cationic water based pigment dispersion[J]. Materials Research Innovations, 121711

17. C. Diaz-Blanco, Gonzalez. M. Diaz, Monmany. J. M. Daga, T. Tzanov, 2009Dyeing properties, synthesis, isolation and characterization of an in situ generated phenolic pigment, covalently bound to cotton Enzyme and Microbial Technology380 EOF385 EOF

18. J. Dong, S. Chen, D. S. Corti, E. I. Franses, Y. Zhao, H. T. Ng, E. Hanson, 2011Effect of Triton X-100 on the stability of aqueous dispersions of copper phthalocyanine pigment nanoparticles[J]. J Colloid Interface Sci, 36213341

19. K. Huhnerfuss, Bohlen. A. von, D. Kurth, 2006Characterization of pigments and colors used in ancient Egyptian boat models[J]. Spectrochimica Acta Part B-Atomic Spectroscopy, 6110-11 12241228

20. S. Leelajariyakul, H. Noguchi, S. Kiatkamjornwong, 2008Surface-modified and micro-encapsulated pigmented inks for ink jet

printing on textile fabrics[J]. Progress in Organic Coatings, 622145161

21. K. Fang, X. Jiang, C. Wang, M. Wu, Y. Yan, 2008Properties of the Nanoscale Hydrophilic Cationic Pigment Based on Quaternary Surfactant[J]. J Disper Sci Technol, 2915257

22. H. Wu, G. Gao, Y. Zhang, S. Guo, 2012Coating organic pigment particles with hydrous alumina through direct precipitation[J]. Dyes and Pigments, 921548553

23. Chang, C., Chang, S., Tsou, S., Chen, S., Wu, F., & Hsu, M.2003Effects of polymeric dispersants and surfactants on the dispersing stability and high-speed-jetting properties of aqueous-pigment-based ink-jet inks[J]. J Polym Sci Pt B-Polym Phys, 411619091920

24. C. Wang, Y. Yin, X. Wang, G. Bu, 2008Improving the Color Yield of Ultra-fine Pigment Printing on Cotton Fabric[J]. AATCC Review 8124145

25. C. Yoon, J. Choi, 2008Syntheses of polymeric dispersants for pigmented ink-jet inks[J]. Coloration Technology, 1246355363

26. N. Bulychev, E. Kisterev, Y. Ioni, O. Confortini, Prez. F. Du, V. Zubov, C. Eisenbach, 2011Sufface Modification in Aqueous Dispersions with Thermo-Responsive Poly (Methylvinylether) Copolymers in Combination with Ultrasonic Treatment[J]. Chemistry & Chemical Technology, 515965

27. S. Fu, C. Du, K. Zhang, C. Wang, 2011Colloidal properties of copolymer-encapsulated and surface-modified pigment dispersion and its application in inkjet printing inks[J]. J Appl Polym Sci, 1191371376

28. Y. M. Chen, R. S. Hsu, H. C. Lin, S. J. Chang, S. C. Chen, J. J. Lin, 2009Synthesis of acrylic copolymers consisting of multiple amine pendants for dispersing pigment[J]. J Colloid Interface Sci, 33414249

29. W. Zhang, S. Deodhar, D. G. Yao, 2010Processing Properties of Polypropylene With a Minor Addition of Silicone Oil[J]. Polymer Engineering and Science, 50713401349

30. J. Wu, L. M. Wang, P. Zhao, F. Wang, G. Wang, 2008A new type of quaternary ammonium salt containing siloxane group and used as favorable dispersant in the surface treatment of CI pigment red 170[J]. Progress in Organic Coatings, 632189194

31. U. Weber, 2010The Effect of Grinding Media Performance on Milling a Water-Based Color Pigment[J]. Chem Eng Technol, 33914561463

32. De Vallejuelo, S., Barrena, A., Arana, G., De Diego, A., & Madariaga, J.2009Ultrasound energy focused in a glass probe: An approach to the simultaneous and fast extraction of trace elements from sediments[J]. Talanta, 802434439

33. X. Jia, C. Caroli, B. Velicky, 1999Ultrasound propagation in externally stressed granular media[J]. Physical Review Letters, 82918631866

34. Johnson, J., Barbato, M., Hopkins, S., & O'Malley, M.2003Dispersion and film properties of carbon nanofiber pigmented conductive coatings[J]. Progress in Organic Coatings, 473198206

35. Y. Yin, C. Wang, Y. Wang, 2012Preparation and colloidal dispersion behaviors of silica sol doped with organic pigment[J]. Journal of Sol-Gel Science and Technology, 62217

36. Fu, S., Xu, C., Du, C., Tian, A., & Zhang, M.2011Encapsulation of CI Pigment blue 15:3 using a polymerizable dispersant via emulsion polymerization[J]. Colloid Surf A-Physicochem Eng Asp, 3841-3 6874

37. Fu, S., & Fang, K.2008Properties of waterborne nanoscale pigment red 122 dispersion prepared by phase separation method[J]. J Appl Polym Sci, 108639683972

38. Squires, T., & Quake, S.2005Microfluidics: Fluid physics at the nanoliter scale[J]. Reviews of modern physics, 773977

39. J. Merrington, P. Hodge, S. Yeates, 2006A High-Throughput Method for Determining the Stability of Pigment Dispersions[J]. Macromolecular Rapid Communications, 2711835840

40. Chang, C., Chang, S., Shih, K., & Pan, F.2005Improving mechanical properties and chemical resistance of ink-jet printer color filter by using diblock polymeric dispersants[J]. J Polym Sci Pt B-Polym Phys, 432233373353

41. F. Nsib, N. Ayed, Y. Chevalier, 2007Selection of dispersants for the dispersion of CI Pigment Violet 23 in organic medium[J]. Dyes Pigment, 741133140

42. D. Xu, K. Fang, S. Fu, 2009Effects of Ethanol on the Stability of Pigment Colloidal Dispersion[J]. Journal of Dispersion Science and Technology, 304510513

43. Wu, H., Huang, K., & Lee, K.2011Supercritical fluid-assisted dispersion of CI pigment violet 23 in an organic medium[J]. Powder Technology, 2063322326

44. P. Karrer, Improvements in or relating to the treatment of cotton fibers preparatory to dyeing. GB Patent 2498421926.

45. Guthrie, J.1947Introduction of Amino Groups into Cotton Fabric by Use of 2-Aminoethylsulfuric Acid [J]. Textile Research Journal, 1711625629

46. H. Wang, D. Lewis, 2002Chemical modification of cotton to improve fibre dyeability[J]. Coloration Technology, 1184159168

47. S. Burkinshaw, X. Lei, D. Lewis, 1989Modification of cotton to improve its dyeability. Part 1-pretreating cotton with reactive polyamide-epichlorohydrin resin[JJournal of the Society of Dyers and Colourists10511391398

48. Hauser, P., & Tabba, A.2001Improving the environmental and economic aspects of cotton dyeing using a cationised cotton[J]. Color Technol, 1175282288

49. H. Jin, 2009The Synthesis and Application of Cationic Polymer (PDA) on Fibers Modification[D]. Master thesis. Jiangnan University;

50. Dubas, S., Chutchawatkulchai, E., Egkasit, S., Lamsamai, C., & Potiyaraj, P.2007Deposition of polyelectrolyte multilayers to improve the color fastness of silk[J]. Textile Research Journal, 776437441

51. J. Wei, 2003Studies on Weighting Methods and Properties of Silk Fiber[D]. Master thesis. Suzhou University;

52. Qin, C., Chen, D., Tang, R., Huang, Y., Wang, X., & Chen, G.2010Studies on silk fabric dyed with a hemicyanine dye[J]. Coloration Technology, 1265303307

53. C. Wang, K. Fang, W. Ji, 2007Superfine pigment dyeing of silk fabric by exhaust process[J]. Fiber Polym, 82225229

54. Murphy, A., John, P., & Kaplan, D.2008Modification of silk fibroin using diazonium coupling chemistry and the effects on hMSC proliferation and differentiation[J]. Biomaterials, 291928292838

55. C. Wang, W. Ji, 2006Dyeing property of the modified wool fabric with superfine pigment[J]. Journal of Textile Research, 2767477

56. X. Yuan, K. J. Fang, 2006Preparation of Waterborne Nanoscale Carbon Black Dispersion with Sodium Carboxymethyl Cellulose[J]. Journal Of Donghua University(English Edition) 235119121

57. R. Yang, M. Dai, 1996Study on Properties of Acid-Modified Multicomponent Copolyesterification Fibre[J]. Journal Of Textile Research, 1712731

58. Y. Zhou, D. Yu, P. Xi, S. L. Chen, 2003Influence of styrene-maleic anhydride copolymers on the stability of quinacridone red pigment suspensions[J]. Journal of Dispersion Science and Technology, 245731737

Index